Near the Horizon

An Invitation to Geometric Optics

Library of Congress Catalog Card Number 2017931118

Print ISBN 978-0-88385-142-5

Electronic ISBN 978-1-61444-030-7

Printed in the United States of America

Current Printing (last digit):
10 9 8 7 6 5 4 3 2 1

The Carus Mathematical Monographs

Number Thirty-Three

Near the Horizon

An Invitation to Geometric Optics

Henk Broer
University of Groningen

Published and Distributed by
THE MATHEMATICAL ASSOCIATION OF AMERICA

THE
CARUS MATHEMATICAL MONOGRAPHS

Published by

THE MATHEMATICAL ASSOCIATION OF AMERICA

The Carus Mathematical Monographs are an expression of the desire of Mrs. Mary Hegeler Carus and of her son, Dr. Edward H. Carus, to contribute to the dissemination of mathematical knowledge by making accessible a series of expository presentations of the best thoughts and keenest research in pure and applied mathematics. The publication of the first four of these monographs was made possible by Mrs. Carus as sole trustee of the Edward C. Hegeler Trust Fund. The sales from these have resulted in the Carus Monograph Fund, and the Mathematical Association of America has used this as a revolving book fund to publish the succeeding monographs.

The expositions of mathematical subjects that the monographs contain are set forth in a manner comprehensible not only to teachers and students specializing in mathematics, but also to scientific workers in other fields. More generally, the monographs are intended for the wide circle of thoughtful people familiar with basic graduate or advanced undergraduate mathematics encountered in the study of mathematics itself or in the context of related disciplines who wish to extend their knowledge without prolonged and critical study of the mathematical journals and treatises.

The following Monographs have been published:

1. *Calculus of Variations,* by G. A. Bliss (out of print)
2. *Analytic Functions of a Complex Variable,* by D. R. Curtiss (out of print)
3. *Mathematical Statistics,* by H. L. Rietz (out of print)
4. *Projective Geometry,* by J. W. Young (out of print)
5. *A History of Mathematics in America before 1900,* by D. E. Smith and Jekuthiel Ginsburg (out of print)
6. *Fourier Series and Orthogonal Polynomials,* by Dunham Jackson (out of print)
7. *Vectors and Matrices,* by C. C. MacDuffee (out of print)
8. *Rings and Ideals,* by N. H. McCoy (out of print)
9. *The Theory of Algebraic Numbers,* second edition, by Harry Pollard and Harold G. Diamond
10. *The Arithmetic Theory of Quadratic Forms,* by B. W. Jones (out of print)
11. *Irrational Numbers,* by Ivan Niven
12. *Statistical Independence in Probability, Analysis and Number Theory,* by Mark Kac
13. *A Primer of Real Functions,* third edition, by Ralph P. Boas, Jr.

14. *Combinatorial Mathematics,* by Herbert J. Ryser
15. *Noncommutative Rings,* by I. N. Herstein (out of print)
16. *Dedekind Sums,* by Hans Rademacher and Emil Grosswald
17. *The Schwarz Function and its Applications,* by Philip J. Davis
18. *Celestial Mechanics,* by Harry Pollard
19. *Field Theory and its Classical Problems,* by Charles Robert Hadlock
20. *The Generalized Riemann Integral,* by Robert M. McLeod
21. *From Error-Correcting Codes through Sphere Packings to Simple Groups,* by Thomas M. Thompson
22. *Random Walks and Electric Networks,* by Peter G. Doyle and J. Laurie Snell
23. *Complex Analysis: The Geometric Viewpoint,* by Steven G. Krantz
24. *Knot Theory,* by Charles Livingston
25. *Algebra and Tiling: Homomorphisms in the Service of Geometry,* by Sherman Stein and Sándor Szabó
26. *The Sensual (Quadratic) Form,* by John H. Conway assisted by Francis Y. C. Fung
27. *A Panorama of Harmonic Analysis,* by Steven G. Krantz
28. *Inequalities from Complex Analysis,* John P. D'Angelo
29. *Ergodic Theory of Numbers,* Karma Dajani and Cor Kraaikamp
30. *A Tour through Mathematical Logic,* Robert S. Wolf
31. *Randomness and Recurrence in Dynamical Systems: A Real Analysis Approach,* Rodney Nillsen
32. *Linear Inverse Problems and Tikhonov Regularization,* Mark S. Gockenbach
33. *Near the Horizon: an invitation to geometric optics,* Henk Broer

MAA Service Center
P.O. Box 91112
Washington, DC 20090-1112
1-800-331-1MAA FAX: 1-301-206-9789

Contents

II Light rays as geodesics

Preface

When you look at the setting sun, sometimes you see surprising things. For instance, occasionally it may look as if the sun is already setting while still above the horizon. Some of you may have witnessed a blank strip in the solar disk, just before sunset. It turns out that these phenomena are related to more familiar things, such as reflection in warm asphalt road paving, or more more generally to mirages and fata morganas. As an example see the photo on the next page, taken by the Vatican Observatory.

One important reason for these phenomena is the following. Normally the temperature decreases with increasing height. Occasionally however, *temperature inversion* occurs in which case a warmer layer may arise above the observer. At the lower boundary of such a layer the light may be either bent (refracted) or bounced (reflected). The latter is called upper reflection. A lower reflection may also occur, in a warm layer near the surface of the earth or near sea level. The latter phenomenon also occurs in smooth sheets of ice, in mirroring lakes, or in the warm asphalt pavings mentioned before. Related to all this is the *green flash* that may arise from the fact that red light rays in the atmosphere are bent more strongly than green ones. As a consequence the "green sun" stays visible, while the "red sun" has already set. Note that usually we observe a white sun as a superposition of the green and the red one. Needless to say that all of this only happens in optimal atmospheric circumstances!

The theoretical background that we are going to develop is called *geometric optics*, that is the theory of light rays. The central idea is the Fermat Principle, which asserts that light rays follows paths of *shortest time*.

PLATE I 3

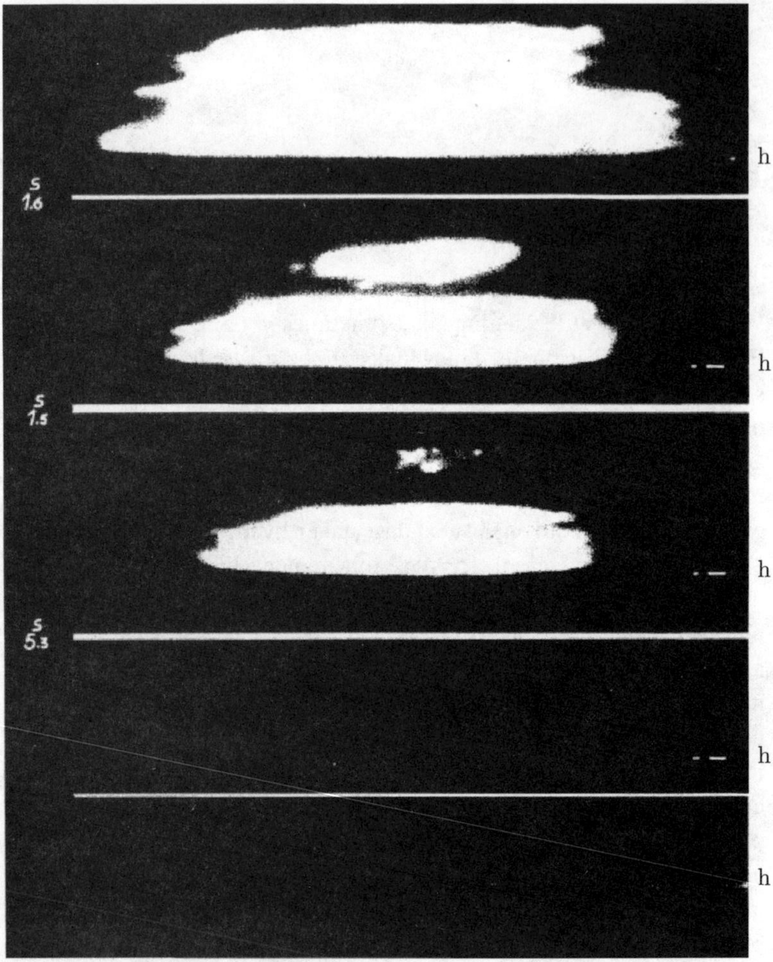

We will also consider the *brachistochrone* problem as this was posed by Johann Bernoulli; this occurred in his Groningen period, between 1695 and 1705. The question is what is the "fastest slideway" between two points in a vertical plane, subject to constant vertical gravitation. This minimization problem is nowadays an exercise in courses on calculus of variations, but Bernoulli's own solution however views the brachistochrone as a light ray in an atmosphere of a flat earth, where the refraction index is properly chosen. He therefore applies the optical Fermat Principle in a mechanical context. This handsome combination will lead us to the famous answer: the best path is a cycloid. We could not resist the temptation to also pay some attention to Christiaan Huygens's earlier work on the cycloid. Thus it will appear that the cycloid, apart from brachistochrone, moreover is *iso-* and *tautochrone*.

A strong motivation and inspiration for the present work came from *The Nature of Light and Colour in the Open Air* by Marcel G.J. Minnaert (1893–1970). This book is the first part of his three volumes *De natuurkunde van 't vrije veld* [72]. A reprint in the 1970s induced a strong revival of interest in Minnaert's work. In the beginning of the 1980s Marja Bos graduated from the University of Groningen with a thesis on *De blinde strook*, i.e., *The blank strip* [19], a theme that will recur below. Minnaert was well-known as an astrophysicist and as the director of the Utrecht observatory Sonnenborgh. For a highly readable biography see Molenaar [73].

The present book is intended for all readers with an interest in the subject, where we especially think of teachers and professors of mathematics and physics. The contents offer many illustrative examples to be used in mathematics or physics classes and may possibly inspire projects, essays, and theses. Therefore, this text may well be of interest to students of these subjects. The book, apart from an introductory chapter, consists of two parts. The mathematics of Part I, entitled *Geometry of light rays in the atmosphere*, mainly consists of planar Euclidean geometry and elementary differential and integral calculus. This material in principle already suffices to support the optical phenomenology of interest to us. Part II, entitled *Light rays as geodesics*, goes deeper into the background mathematics in terms of the calculus of variations and differential geometry, where the Fermat Principle is turned into Hamilton's Principle. Here a magnificent background culture is touched on that makes a connection with many later developments of mathematical physics. The chapters, as well as the entire book, are concluded with a *scholium* where suggestions for further reading are collected. An earlier version of the present book appeared in Dutch [24].

Acknowledgement. During the preparation of the book in hand I enjoyed inspiring discussions with a large number of people. Particularly my gratitude goes to Jan Aarts, Frits Beukers, Marja Bos, Konstantinos Efstathiou, Aernout van Enter, Wout de Goede, Jan Guichelaar, George Huitema, Hildeberto Jardón-Kojakhmetov, Jan van Maanen, Sjoerd Rienstra, Jan Scholtmeijer, Henk de Snoo, Alef Sterk, Floris Takens, Gert Vegter, Ferdinand Verhulst, Holger Waalkens, and Florian Wagener. Steven Duintjer Tebbens is thanked for a final, careful check of the text, and so is Fernando Gouvea from the MAA: both made valuable suggestions for improvement. Konstantinos and Hildeberto are acknowledged for their tremendous help in developing the necessary numerical algorithms and the preparation of the many figures.

Henk Broer, Groningen, Spring 2015

Chapter 1

Introduction and Overview

This book is concerned with certain near-horizon optical phenomena, such deformations of the solar disk just before sunset and fata morganas and mirages, as already discussed in the Preface. In particular, our interest is in the development and application of geometric optics in an atmospheric context. We will study how light rays are refracted and reflected through layers in the atmosphere. This leads to an explanation of the various optical phenomena that may occur around the horizon. An important example of this is formed by blank strips in the disk of the setting sun, as described by Wegener [93, 94] and later by Minnaert [72]. For a schematic impression, see Figure 1.1 on page 2, which we now inspect in more detail.

Usually temperature decreases monotonically with the distance to the earth's surface, but sometimes a *temperature inversion* occurs, in which one or more warmer layers come into existence. Since the refraction index depends on the temperature, an inversion can create a discontinuity in the refraction index profile. In Figure 1.1 the situation is depicted with one warm layer bounded from below by a circle at height H that resides above the observer W.[1] In the circle at height H reflection can occur, often indicated by the term *upper reflection*. As we will explain, in interaction with refraction a *Wegener sector* can arise, within which W only receives light rays from below the horizon. When the sun is at the appropriate

[1] The Dutch name for observer is "waarnemer", which explains the notation W.

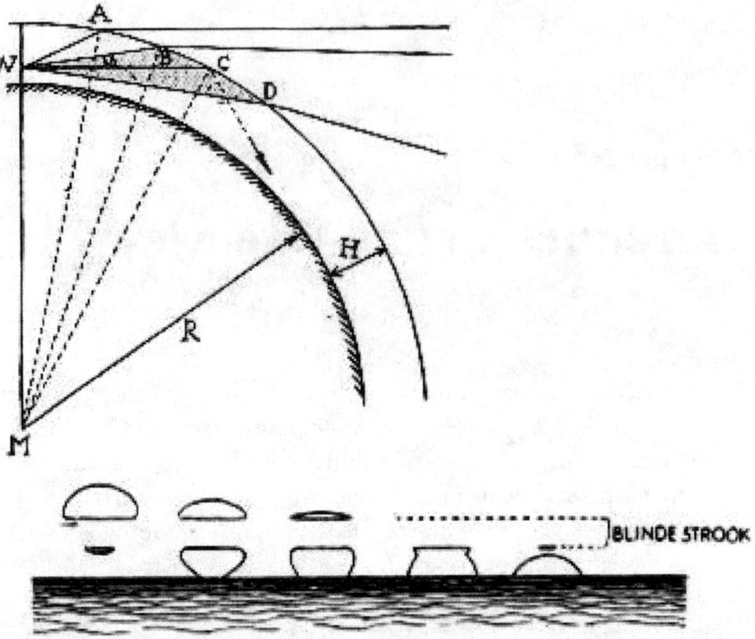

Figure 1.1. Minnaert [72], Figure 60. Qualitative explanation for a blank strip (Dutch: *blinde strook*) in the sun in a model with one warm layer. The lower boundary of the layer is formed by a circle at height H. The observer W roughly looks in the horizontal direction. Transitions between refraction and reflection yield a so-called *Wegener sector*, within which W only receives light rays from below the horizon. This may lead to a blank strip in the setting sun. See the text for further explanation.

height above the horizon, W observes a blank strip in the solar disk, also see Figure 1.2 on page 3.

Furthermore, within or near such a Wegener sector W may witness fata morganas: these are images of objects below the horizon. In Chapter 3 we will discuss this in more detail and also discuss variations where lower reflection occurs, for instance in a warm layer close to the earth's surface. As an example see Figure 1.4 on page 6.

These descriptions have a largely schematic nature: they serve as a qualitative support for understanding rather than for strict quantitative

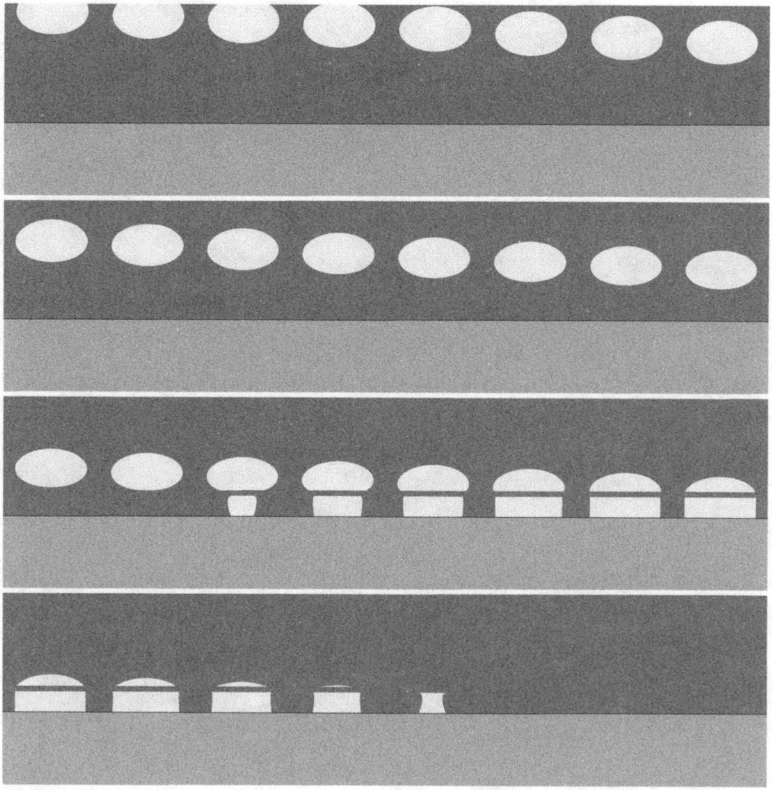

Figure 1.2. Theoretical sunset with blank strip. The atmosphere is dark and the earth light gray. Also note the deformation of the solar disk. For details see Chapter 3.

purposes. This is a tradition that goes back at least as far as Kepler [57]. Precise data of refraction index profiles over the entire height of the atmosphere are in fact not available. For the (considerably awkward) physical background see, for instance, the Feynman Lectures [38]. In fact we here are dealing with a so-called *inverse* problem, namely to what extent certain phenomena in the atmosphere can be given a geometric optical explanation; see also [65, 95, 96].

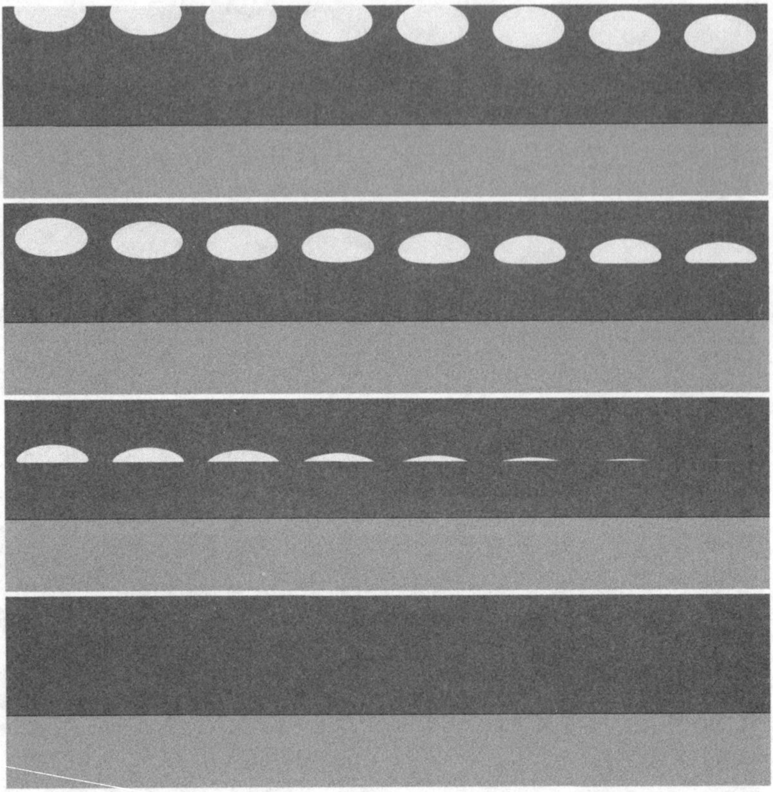

Figure 1.3. Theoretical sunset with a blank zone. It now appears as if the sun already sets above the horizon. This is a quite common phenomenon. For details again see Chapter 3.

Remarks.

– Some relevant data. The diameter of the sun is approximately $32' \approx 0.5^\circ$, expressed in minutes and degrees of arc. The radius of the earth is $R = 6400$ km, while the height of the atmosphere approximately equals 11 km. Realistic values for the height of the observer W are 10 to 15 m, if he observes from the top of a dune or from the deck of a ship.

A "warm layer" might be only a few degrees warmer, but sometimes the difference is as much as 20° Celsius.

- In a model with one warm layer as in Figure 1.1, a value of $H \approx 80$ m as the height of the lower circular boundary of the warm layer is often appropriate for "saving the appearances".[2] For somewhat more realistic models see Chapter 3.
- The sunsets and fata morganas depicted here and in Chapter 3 are referred to as "theoretical". This means that the pictures are produced by simulation based on the geometric optical theory developed in this book.
- For the sake of clarity the more schematic figures, such as Figure 1.1 and also a few in Chapter 3, have deliberately been depicted far beyond their realistic proportions.

The present chapter forms an introduction to, and simultaneously a summary of, the background geometic optical theory, as this is spread out over the following chapters. This is why at this moment not all details will be spelled out, but we refer to later chapters instead.

Alfred L. Wegener (1880–1930) was a versatile scientist who became most renowned for his theory of floating continents, which was accepted by the geological community only 30 year after his death. His qualitative approach for atmospheric-optical phenomena has a central role in the present book. This mainly concerns the Wegener sector, that will be detailed in Chapter 3.

1.1 Towards a mathematical description

Our mathematical description of the phenomena described above rests on the Fermat Principle, which asserts that light rays *by definition* follow

[2]This expression goes back on Plato and Aristoteles, compare Dijksterhuis [33]. See our discussion on the inverse problem.

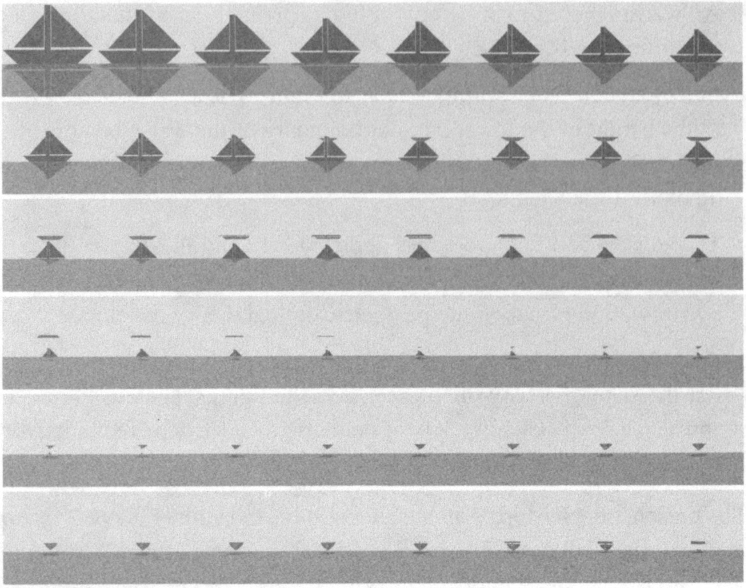

Figure 1.4. Theoretical fata morganas of a sailing ship disappearing slowly below the horizon. In this case both upper and lower reflections occur. For details again see Chapter 3.

paths of shortest time [5, 70, 83]. For an explanation how this principle is connected with wave optics we refer to Feynman et al. [38] or to Arnold [7].

Below we shall regularly use the terms *isotropic* and *homogeneous* as possible properties of an optical medium. Isotropy means that in each point of a medium the speed of light is equal in all directions, say with magnitude v. This justifies a pointwise definition of the refraction index by $n = c/v$, where c is the speed of light in vacuo. Homogeneity means that v, and therefore n, is independent of the position in the medium. The Fermat Principle implies that in a homogeneous medium the light rays form straight lines that are traversed with a constant velocity. If the medium consists of two homogeneous layers bounded by a flat plane, then in that plane reflection or refraction takes place according to certain laws that will be spelled out in the next chapter.

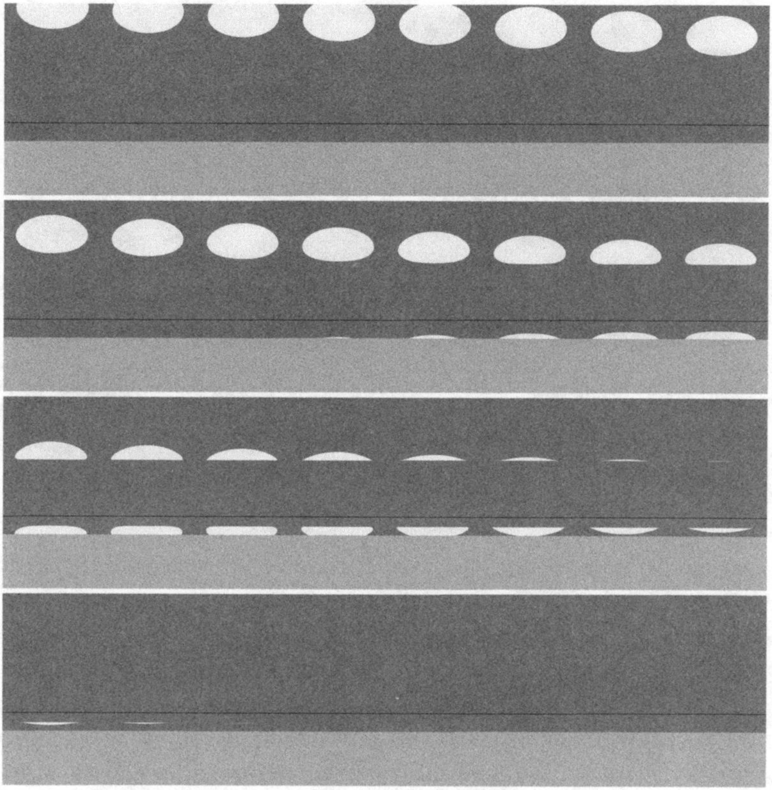

Figure 1.5. Theoretical sunset with blank zone as can occur in the presence of a Wegener sector together with lower reflection. Note that the latter images of this series reproduce images of pieces of the upper solar disk by lower reflection, that W observes upside down. For more details see Chapter 3.

1.1.1 Theoretical assumptions

Our approach is mainly restricted to one vertical plane passing through the eye of the observer and one other point, say, the center of the solar disk, the moon disk, a planet, a ship or mountain, etc.

In the vertical plane we consider, two natural coordinate systems are used, depending on whether we work either with a *flat* or a *round* earth.

Figure 1.6. Minnaert [72], Figure 62: two single snapshots of moon sickles that are optically deformed by the atmosphere. Such phenomena cannot be explained by a refraction index profile that only depends on the vertical height. A description in terms of one vertical plane through the eye of the observer no longer suffices, in fact.

For the flat earth this system is cartesian, where x is the horizontal and y the vertical coordinate. For the round earth we use polar coordinates φ and r with respect to the center of the earth as horizontal and vertical coordinates, respectively. Here r is the distance to the center and φ the angle with respect to a fixed direction. Of course the earth is *not* flat, but it is certainly possible to consider the resulting model. For one thing, it is easier to understand. Surprisingly, it will turn out to be relevant to the brachistochrone problem, see Chapter 4.

Symmetry. The layers of different refraction index in the case of a flat earth will be assumed to be horizontal, i.e., invariant under horizontal translations. In case of a round earth, we analogously assume that these layers are invariant under rotations about the center of the earth. In other words this means that the refraction index n does not depend on either x or on φ. In the former case we represent it as a function $n = n(y)$ and in the latter as $n = n(r)$.

Remarks.

– That the refraction index is independent of x or φ expresses a symmetry property of the atmosphere. We like to note that for our considerations this symmetry only has to be valid for a relevant sector of the atmosphere, usually a small neighborhood of the horizon.

– Evidently rotational symmetry is not always realistic, for instance see Figure 1.6 on page 8; for an explanation of such a phenomenon even the restriction to one vertical plane is unrealistic. In Chapter 3 we shall encounter the Nova Zembla phenomenon. Its decription also requires a refraction index profile that is not rotation invariant. In Chapter 5 we present the mathematical theory needed for the computation of general light rays in three–dimensional space. See Chapter 8 for optical remarks that are beyond the scope of this book.

1.1.2 Discrete versus continous atmosphere, conservation laws

Often it is convenient to work with *discrete* homogeneous layers of varying refraction index. At the boundaries between layers the refraction and reflection laws are valid as mentioned before and the light ray follows a broken line in the atmosphere. Both for refraction and for reflection a formula

$$n_2 \sin \alpha_2 = n_1 \sin \alpha_1$$

will turn out to hold, where the two bounding layers are numbered 1 and 2 and where α is the angle between the light ray and the normal on the boundary. For the case of refraction this formula is known as Snell's Law, see Figure 1.7. In this book α is always the angle with the vertical direction, to be called *inclination*. The inclination α thus becomes a parameter along the light ray.

In the discrete case $n = n(y)$ or $n = n(r)$ is a discontinuous step function. We also shall consider the case where these functions are continuous or smooth. Note that step functions with an increasing number of steps can be used to approximate the continuous case. It turns out that in all cases this will lead to a *conservation law*. In the flat case the quantity

$$S = n(y) \sin \alpha \tag{1.1}$$

will turn out to be conserved along the entire light ray, while in the round case this holds for

$$C = rn(r) \sin \alpha \,. \tag{1.2}$$

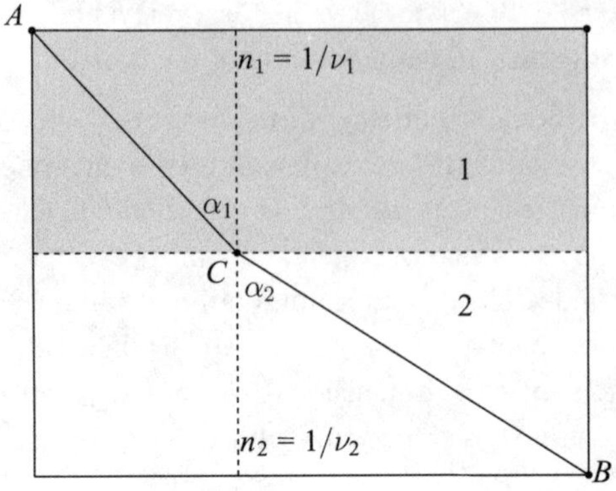

Figure 1.7. Refraction law of Snell. The broken line *ACB* will turn out to be the path of shortest time to travel from *A* to *B* via the boundary between the layers 1 and 2, when $n_2 \sin \alpha_2 = n_1 \sin \alpha_1$.

The conservation laws corresponding to (1.1) and (1.2) directly follow from the reflection and refraction laws mentioned before, using a straightforward application of planar Euclidean geometry. As a consequence of these conservation laws, the refraction index profiles $n = n(y)$ or $n = n(r)$ completely determine the evolution of the inclination $\alpha = \alpha(y)$ or $\alpha = \alpha(r)$ and therefore the position of the light ray in the (x, y)–or the (φ, r)–plane. This consequence plays a key role throughout the book. For a detailed discussion we refer to Chapter 2.

Remarks.

- The equations of $\alpha = \alpha(y)$ and $\alpha = \alpha(r)$ for given profiles $n = n(y)$ and $n = n(r)$ may well contain integrals that cannot be expressed in elementary functions. Still, with numerical methods, given any profile, the light rays can be approximated, even for more general refraction index profiles $n = n(x, y)$ and $n = n(\varphi, r)$, and also when the light rays do not remain in one vertical plane.

- The speed of propagation in this description always is given by $v(y) = 1/n(y)$, or $v(r) = 1/n(r)$, where the speed of light in vacuo is reduced to $c = 1$ by an appropriate choice of units.

1.2 Overview of the contents

Part I of this book is based on the direct use of differential and integral calculus and some Euclidean geometry. Part II is meant to deepen this material in terms of calculus of variations and differential geometry. There the Fermat Principle is turned into that of Hamilton, where we tried to keep the mathematical formalism minimal.

1.2.1 On Part I: Geometry of light rays in the atmosphere

Part I consists of five chapters based on elementary calculus and some planar Euclidean geometry, in particular the law of sines and the fact that the inner and outer bisectrix of any angle are perpendicular. The theory will be developed in Chapter 2 and in Chapter 3 will be applied to light rays in the atmosphere. In particular this allows us to understand phenomena near the horizon.

In Chapter 4 the brachistochrone problem will be discussed. The brachistochrone is the form of a smooth wire profile between two given points A and B in a vertical plane such that a bead K that slides along this profile minimizes time (Greek: *brachistos* = shortest, *chronos* = time), as in Figure 1.8 on page 12. The bead is assumed to be subject to the (constant) gravitational force and the sliding motion takes place without friction. We will see how Johann Bernoulli's solution rests on the mathematics of atmospheric optics on a flat earth, by viewing the brachistochrone curve as a light ray. The refraction index profile $n = n(y)$ then is chosen according to conservation of energy of the motion in the constant gravitational field. With this problem Bernoulli was one of the founding fathers of the calculus of variations and the Variational Principle, which have been of tremendous importance for the development of the sciences since then, see [7, 38, 43, 85, 91]. The brachistochrone curve turns out to be a cycloid, a curve that earlier in the 17th century was identified by Huygens as the iso- and tautochronous curve.[3] Isochrony means that for a sliding

[3] It seems that this form also inspired the design of skate board tracks.

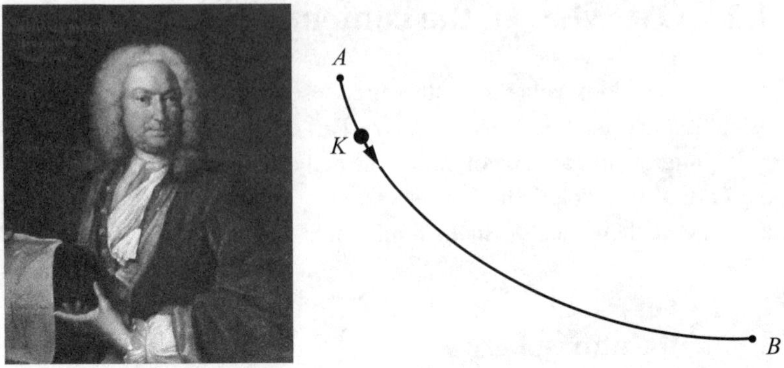

Figure 1.8. Left: Johann Bernoulli (1667–1748) painted by Johann Rudolf Huber, Alte Aula Universität Basel. Right: Illustration of the brachistochrone problem [14], that asks for a wire profile that minimizes the time for a bead K to slide from A to B. The frictionless motion takes place in a constant vertical gravitation field.

bead that oscillates along the cycloid, the period does not depend on the amplitude. Tautochrony says that the time for the bead to fall from any point of the cycloid to its minimum is always the same. (Greek: *isos* = equal, *tautos* = same.) In § 4.2 we shall see how both these properties follow from the fact that the bead, oscillating along the cycloid, performs a harmonic oscillation. See also [24, 27].

The mathematician Johann Bernoulli (1667–1748) was born and died in Basel, Switzerland, where he also lived most of his life. Between 1695 and 1705 he worked in Groningen, where, among other things, he proposed the brachistochrone problem. This marks a starting point of the development of the calculus of variations. His son Daniel was born in Groningen in 1700. Both these Bernoullis were members of a large family, many of whom were active in mathematical and physical subjects, for instance in mechanics and hydraulics. The Bernoulli family were not a very peaceful lot: matters of priority led to big conflicts. For more information we refer the reader to Van Maanen [70]. Johann Bernoulli was the teacher of both his son Daniel and of Leonhard Euler (1707–1783).

Figure 1.9. Alfred Wegener (1880–1930) and Marcel Minnaert (1893–1970).

1.2.2 On Part II: Light rays as geodesics

Chapter 5 will provide a mathematical exploration and deepening of the Fermat Principle, based on methods as developed in the 18th and 19th century by, among others, Euler [36], Lagrange [61] and Hamilton [47]. Elementary differential geometry will also be introduced: light rays turn out to exactly move along geodesics in an appropriate Riemannian metric [82]. In this setup, the calculus of variations will be applied, with the help of the Euler–Lagrange equations and a translation of these that leads to the formulation of Hamilton's Principle [7, 47] for geometrical optics. We will need more advanced mathematical ideas to clarify all this. One of the main tools consists of the following: Next to the *position* of the light ray we will also need to independently take its *velocity* into account.

In Chapter 6 we shall revisit the brachistochrone problem, now armed with the extra mathematics of Chapter 5. As before we find ourselves in the atmosphere of a flat earth with a refraction index profile given by the constant gravitational field. The brachistochrone curve now is rediscovered as a geodesic in an appropriate Riemannian metric, determined by a refraction index $n(y)$ related to the conservation of energy in the constant gravitational field. Moreover it turns out that the reduced problem can be integrated in terms of elementary functions. Here it becomes clear that the inclination α with the vertical direction acts as a geodesic parameter. This solution of the brachistochrone problem runs in a manner that strongly reminds one of the way Kepler's laws follow from Newton's principles, for which see [7].

Chapter 7 discusses a connection between light rays in the rotationally symmetric atmosphere of the round earth and the geodesics on a surface of revolution in three–dimensional space. It turns out that the conservation law regarding the quantity C (in (1.2)) exactly coincides with Clairaut's Theorem [7, 30]. It now becomes an interesting problem to what extent $\mathbb{R}^2 \setminus \{0\} = \{\varphi, r\}$, endowed with the optically relevant Riemannian metric, can be isometrically embedded as a surface of revolution in $\mathbb{R}^3$. It will turn out that this embeddability holds true for a large number of relevant refraction index profiles $n = n(r)$.

Finally in Chapter 8 we present a few miscellaneous remarks regarding optics beyond the scope of the previous chapters. We briefly touch on the concept of caustic, with the rainbow as a particular case. For further details see, e.g., [15, 16, 72, 81].

1.3 Scholium

From the 17th century on mechanics, optics and mathematics have developed simultaneously. This started with Kepler, Galileo, Descartes, Fermat, Huygens, Newton, the Bernoullis, Leibniz, and many others. For handsome overviews see Dijksterhuis [33], Koestler [59], Linton [69], Sabra [83], and Van der Werf [95]. For interesting historical vistas see Arnold [10], H.J.M. Bos [18] and Van Maanen [70]. Much of the mathematics that nowadays belongs to many standard curricula of secondary and higher education had already come into existence at that point.

Evidently this corpus of knowledge and insight has signifcantly expanded and deepened since then.

The renewal mainly consisted of the invention and development of differential and integral calculus and of analytic geometry. In the following two centuries this package was extended by the calculus of variations [7, 42, 44] and differential geometry [30, 88]. The interest in ordinary differential equations was extended to partial differential equations. The above list thereby extended to, among others, Euler, Lagrange, Legendre, Gauss, Hamilton and Riemann.

Large portions of natural philosophy were gradually mathematized during this period, thereby leading to the disciplines of physics, astronomy and, later, also chemistry. Via partial differential equations the dynamics of continua (such as spatial patterns) could be studied, which eventually enabled a mathematical foundation of wave optics. A climax was the electromagnetism of Maxwell, that also contains a wave theory of light [38, 46, 81]. Evidently there is much more. We only need to mention the long and extremely interesting road that led to modern theoretical and mathematical physics and chemistry. This abundance enforces a clear definition of the contents to be discussed in the book at hand, where we restrict ourselves to geometric opics.

In the area of geometric optics there is a lot more to read. Surely the work of Carathéodory [29] and of Berry [15, 16] have to be mentioned. Also see, e.g., Leonhard and Philbin [67]. Also worth reading are the classical Feynman Lectures on Physics [38], in particular part I, § 26, and, somewhat later, Guillemin and Sternberg [46], Ch. III, or Poston and Stewart [81], Ch. 12. For more physical backround on atmospherical optics see, for instance, Humphries [54]. There is a large amount of physical-meteorological literature on the fascinating phenomenology around the setting sun, in recent years increasingly illustrated by fantastic photo and film material. For the latter the reader may consult the internet as well. In this respect we would like to mention Floor [39, 40].

As indicated earlier, in the atmosphere it is virtually impossible to get actual knowledge of the atmospheric refraction index profile $n = n(r)$ or, more generally, $n = n(\varphi, r)$. Armed with the mathematics of geometric optics, we are able to attempt a reconstruction of such a refraction index profile from the observed optical phenomena, and so (perhaps?) to provide a rational explanation for what we see. Before we called this an inverse problem. Lehn [65] and Van der Werf [95] solve similar inverse problems, thereby also giving meteorological and physical-geographical explanations of the optical phenomena they discuss.

This interaction between theory and phenomena, between mathematics and natural science has gone out of use, in recent decades at many secondary schools in the western world. Mathematics and the various natural sciences are being taught almost in complete mutual isolation. In certain cases pious intentions have been expressed regarding modelling and feedback mechanisms between the model and the object of modelling. What regularly fails is the realization that in important parts of the natural sciences models are based on a theory. Such a theory then yields a connected corpus of models. To stay with atmospheric optics, one might say that every individual refraction index profile could be called a model, fitting within the mathematical background of geometrical optics.

The contributions of Minnaert [72], of Van der Werf [95], but also for instance of Levi [68], highlight the connection and harmony of mathematics and the natural sciences. The present book attempts to contribute to this insight as well.

We would like to conclude this chapter by remarking that the mathematics of this book deliberately has been kept anachronistic. For instance, neither Kepler nor Huygens had differential and integral calculus at their disposal, but we however apply them without scruples. We tried to dress everything in a 20th and 21st century coat, for the sake of readability.

I

Geometry of light rays in the atmosphere

Chapter 2

Geometric optics

As we discussed in § 1.1, geometric optics works with light rays that, according to the Fermat Principle (Pierre de Fermat 1607/08–1665), follow paths of shortest time by definition. One also speaks of the "principle of shortest time". We first consider the situation where an optical medium consists of discrete layers, separated by smooth surfaces. As in § 1.1, we assume that the layers are homogeneous, which means that within a layer the light rays follow straight lines with a constant velocity. If this velocity of propagation has value v, the *refraction index* by definition equals $n = c/v$. Therefore one has $v = c/n$. Here c is the velocity of light in a vacuum. To simplify, we choose units so that $c = 1$.

We shall have to distinguish between a real minimum of the time, or only an *extremum*, where it remains to be decided whether this is a minimum, a maximum or yet something different. If the light ray only has to follow an extremal time, we speak of the *weak* Fermat Principle. As we shall see, "locally" the two versions of the principle are equivalent.

Remarks.

- We have to modify slightly our definition of a light ray, which, in fact, has to satisfy only the weak Fermat Principle. As we will see, in many cases the "real" Fermat Principle does apply, but there exist physical examples for which we need the weak version. For instance consider the case where a light source is situated at one of the focal points of an ellipse, where the possible light rays are reflected in the ellipse. Then all possible light rays will pass through the other focal point, because all the travel times are identical.[1] For more details see below.

– Instead of "extremum" one sometimes says "critical point."

On the boundary surfaces both reflection (bouncing) and refraction (bending) may occur. For this we shall first encounter the two classical laws, named after Hero of Alexandria (Hero(n) of Alexandria 10-70) and Snell (Willebrord Snel van Royen 1580-1626), which, in fact, both were already known in ancient times. Below we shall derive both laws from the weak Fermat Principle, as did Leibniz [66] (Gottfried Wilhelm Leibniz 1646–1716).

2.1 More precise setting of the problem

Anticipating our atmospheric applications, the optical considerations will be restricted mainly to a vertical plane that passes through the eye of the observer and one other point determined by our application. That other point could be the center of the sun, the moon or another optical object.

We distinguish between two different situations, namely that of a *flat* and of a *round*[2] earth. The flat option is a nice mathematical simplification that turns out to be useful for the discussion of Bernoulli's brachistochone problem. As announced in Chapter 1, in the vertical plane we use cartesian coordinates (x, y) in the flat earth case and polar coordinates (φ, r) in the round earth case, see Figure 2.1 (page 21). The coordinates x and φ then are horizontal, i.e., parallel to the surface of the earth. Furthermore y and r always are the vertical coordinates, and r is measured in the radial direction, counting from the center of the earth.

In both cases it is assumed that the refraction index n does not depend on the horizontal coordinate, so in the case of the flat earth we have

[1] This is the so-called *gardener property* of the ellipse.

[2] or spherical.

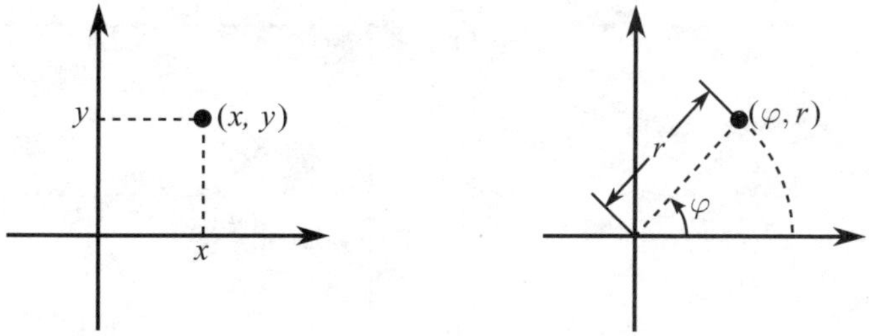

Figure 2.1. The plane with cartesian coordinates (x, y) (left) and polar coordinates (φ, r) (right). We recall the relation $x = r\cos\varphi$ and $y = r\sin\varphi$.

$n = n(y)$ and for the round case $n = n(r)$. These functions will be called *refraction index profiles.*

We shall start with n being a step function and take limits to the case where the refraction index n depends continuously or even smoothly on the position. Here the number of layers tends to infinity, while the thickness of each layer tends to zero. In this way homogeneity is abandoned, though isotropy is being maintained.

Remarks.

- In Chapter 5, the geometric optical theory is developed also for the case where the refraction index profile depends on both variables (x, y) or (φ, r), or even on the full three-dimensional position.
- The theory yields differential equations that lend themselves well to numerical integration and therefore also to create images like the ones used in this book.

2.2 Derivation of the reflection and refraction laws

In this section we shall see how the classical reflection and refraction laws of Hero and Snell can be derived from the weak Fermat Principle. We shall also study the transition between refraction and reflection.

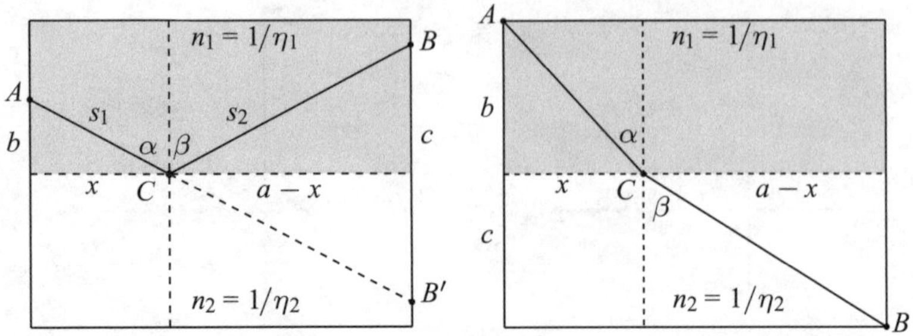

Figure 2.2. The (weak) Fermat Principle leads to reflection in accordance with Hero (left) and refraction in accordance with Snell (right). In both cases the broken line ACB forms the path of shortest time to travel from A to B via the boundary $y = 0$. Notice that B' (left) is the mirror image of B in this boundary.

2.2.1 A straight boundary

In the planar medium we distinguish two layers separated by a horizontal line, say, given by the equation $y = 0$, see Figure 2.2 on page 22. Each layer itself is homogeneous, with constant velocity of propagation v_1 in the upper halfplane $y > 0$ and v_2 in the lower halfplane $y < 0$.

In the plane two points A and B are given, where A lies in the upper halfplane $y > 0$. The point B can lie either in the upper or in the lower halfplane. The question now is how a light ray should go from A to B according to the weak Fermat Principle, when passing through a point C on the boundary $y = 0$. The point C must be therefore be chosen in such a way that the sum of the time intervals t_{AC} and t_{CB}, needed to travel from A to B via C, is extremal.

An important ingredient here is the angle between the light ray and the vertical. In a homogeneous layer this angle does not change and therefore we are only interested in what happens with this angle on the boundary line. Let α and β be the (acute) angles of the incoming and outgoing light ray in the desired point C, see Figure 2.2.

Theorem 2.1 (Laws of Hero and Snell). *Assume the above circumstances, where A lies in the upper halfplane $y > 0$ and C on the boundary $y = 0$. Then the broken straight line ACB is a light ray according to the weak Fermat Principle if and only if the following holds.*

1. In case B is also in the upper halfplane $y > 0$ *one has*

$$\beta = \alpha \quad \text{(reflection)};$$

2. In case B is in lower halfplane $y < 0$ *one has*

$$n_2 \sin\beta = n_1 \sin\alpha \quad \text{(refraction)}.$$

Proof. The proof of Theorem 2.1 consists of a few explicit computations.

Let $x = x_C$ indicate the position of C on the line $y = 0$ that separates both layers. Let t_{AC} be the time needed to travel from A to C and, likewise, t_{CB} the time needed to travel from C to B. We then have to find an extremum of $t_{AC} + t_{CB}$. The facts that $|A - C| = t_{AC} \times v_1$ and that $v_1 = 1/n_1$ yield

$$t_{AC} = n_1 \, |A - C|.$$

By the Pythagorean Theorem we know that

$$|A - C| = \sqrt{x^2 + b^2} \quad \text{so} \quad t_{AC} = n_1\sqrt{x^2 + b^2}.$$

Differentiating t_{AC} with respect to x then leads to

$$\frac{d}{dx} t_{AC}(x) = \frac{n_1 x}{\sqrt{x^2 + b^2}} = n_1 \sin\alpha.$$

Analogously we find

$$\frac{d}{dx} t_{CB}(x) = \begin{cases} -\dfrac{n_1(a-x)}{\sqrt{(a-x)^2 + c^2}} = -n_1 \sin\beta & \text{when } B \text{ in upper halfplane} \\ -\dfrac{n_2(a-x)}{\sqrt{(a-x)^2 + c^2}} = -n_2 \sin\beta & \text{when } B \text{ in lower halfplane,} \end{cases}$$

again see Figure 2.2, left and right. We conclude that

$$\frac{d}{dx}(t_{AC} + t_{CB})(x) = 0 \tag{2.1}$$

if and only if

$$\begin{array}{lll} \alpha = \beta & \text{in case } B & \text{in upper halfplane} \\ n_1 \sin\alpha = n_2 \sin\beta & \text{in case} B & \text{in lower halfplane} \end{array}$$

where in the former case we need to note that $n_1 \sin\alpha = n_1 \sin\beta$ is equivalent to $\alpha = \beta$, since both angles are acute. □

Remarks.

- In the case of reflection ACB' is a straight line, where B' is the mirror image of B in the boundary line $y = 0$. Clearly this is the shortest path in distance, which now also corresponds to shortest time.
- Note that both for refraction and reflection the following formula holds true:

$$\frac{\sin\alpha}{\sin\beta} = \frac{n_2}{n_1} = \frac{v_1}{v_2}.$$

- The formula (2.1) only implies that the time $t_{AC} + t_{CB}$ needed is extremal, but does not tell us the nature of the extremum. To find this out we also have to determine the second derivative of $t_{AC} + t_{CB}$, which after some calculations turns out be positive, meaning that the extremum is indeed a minimum. This implies that in this case both versions of the Fermat Principle are equivalent.

2.2.2 A curved boundary

If the boundary between layers is a smooth curve, virtually the same considerations apply, where in the analogue of Theorem 2.1 the angles α and β have to be measured with respect to the normal to the tangent line in C. However, here minimality of the travel time only holds locally, since further away from the curve other situations may occur as will be explained by the following example.

We already saw the example where the boundary has the form of an ellipse and A and B are the focal points. In that case any point C on the ellipse leads to an extremum, since now $t_{AC} + t_{CB}$ is *constant*, i.e., independent of C, for instance see [7, 38, 46]. The observer therefore should see light coming from all directions, which also reflects the origin of the term "focal point". This is a quite extreme example of a *caustic*. In cases where the extremum is less degenerate the caustics that appear "generically" are related to Singularity Theory, compare Chapter 8 and, for instance, [81].

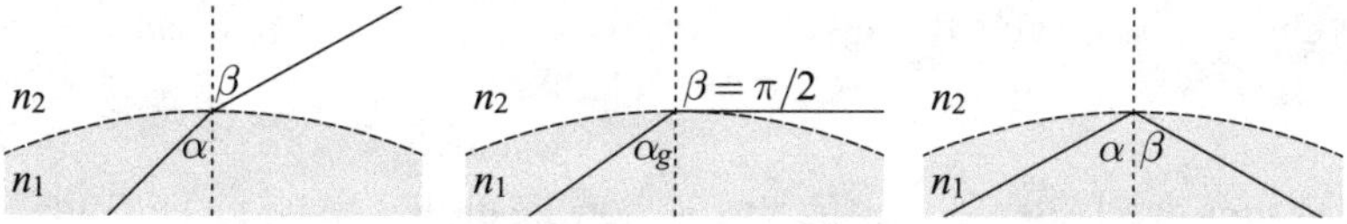

Figure 2.3. Refraction (left) and reflection (right) in a circular boundary; both cases are separated by the limit angle (middle). In this figure $n_1 > n_2$, but now the rays come from below. The reason for this reversed representation will be made clear in Chapter 3.

In what follows these considerations are used for a circular boundary. This is inspired by the atmospheric application, where the spherical symmetry of the earthly atmosphere in a vertical plane gives rise to concentric layers with circular boundaries. See Figure 2.3.

2.2.3 The limit angle

We assume that $n_1 > n_2$ and consider Figure 2.2 from a somewhat different viewpoint, namely where the point C on the boundary is kept fixed, while the light ray enters from the upper halfplane $y > 0$, where the angle α with the y–direction varies from 0 to $\frac{1}{2}\pi$. Our interest then is how β will depend on α. To begin with, we have $\alpha = 0$, in which case the broken light ray passes vertically through the boundary $y = 0$, compare Figure 2.2 (page 22), right. Initially also the exit angle β increases with α. However, for a certain value $\alpha = \alpha_g < \frac{1}{2}\pi$ the angle β reaches the value $\beta = \frac{1}{2}\pi$. To be precise, this happens at

$$\alpha_g = \arcsin\left(\frac{n_2}{n_1}\right); \tag{2.2}$$

this is the so-called *limit angle*, a concept that goes back to Wegener [93, 94]. At that moment the outgoing light ray is *tangent* to the x–axis. When α increases further the light ray bounces back into the upper halfplane $y > 0$, as in Figure 2.2, left.

One might say that at $\alpha = \alpha_g$ the exit angle β has a discontinuity and jumps from $\beta = \frac{1}{2}\pi$ to $\beta = \pi - \alpha_g$. From that moment on β again changes continuously with α and decreases until $\beta = \frac{1}{2}\pi$, which occurs for $\alpha = \frac{1}{2}\pi$.

A notation issue. When trying to reconcile the left and right half of Figure 2.2 a notation issue shows up. To resolve this we now introduce next to the entrance angle α an exit angle α', as measured from the negative y–axis. In the case of refraction α' is acute while α' is obtuse. In terms of Figure 2.2 this means

$$\alpha' = \begin{cases} \beta & \text{in the right picture (refraction)} \\ \pi - \beta & \text{in the left picture (reflection),} \end{cases} \tag{2.3}$$

see Figure 2.4. In Figure 2.5 we depict α' as a function of α for for $0 \le \alpha \le \frac{1}{2}\pi$. The discontinuity at the limit angle $\alpha = \alpha_g$ now is clearly visible.

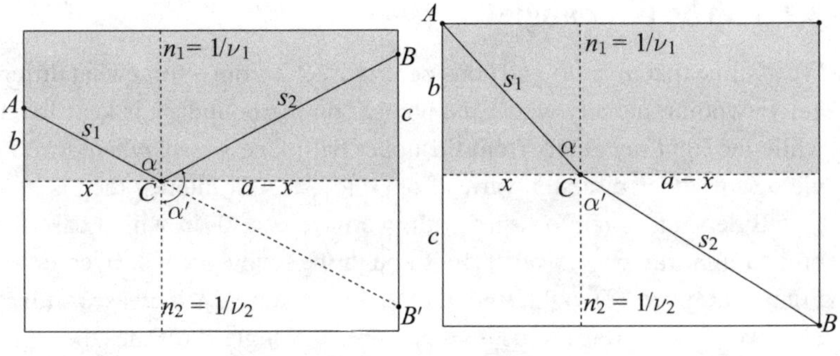

Figure 2.4. This version of Figure 2.2 indicates the angles α and α', as specified in formula (2.3).

Reformulation of Hero and Snell. In terms of the notation α, α' Theorem 2.1 has the form

$$n_1 \sin\alpha = n_2 \sin\alpha', \tag{2.4}$$

where the difference between refraction and reflection only is expressed by the fact that α' is acute or obtuse and whether or not $n_1 = n_2$. In fact the sine of the angle the light ray makes with the vertical is more important

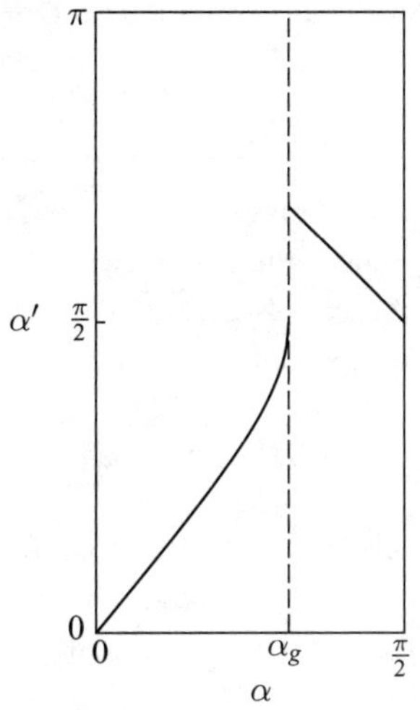

Figure 2.5. Graph of α' against α for $0 \leq \alpha \leq \frac{1}{2}\pi$, compare Figures 2.2 and 2.4. At the *limit angle* $\alpha = \alpha_g$, see (2.2), a discontinuity shows up and refraction jumps into reflection.

than the angle itself. As we shall see, (2.4) is a convenient reformulation of Theorem 2.1 for the case where the light ray passes through several parallel layers. Whenever appropriate we use, of course, that $\sin(\pi - x) = \sin x$.

2.3 Conservation laws

In atmospheric applications, besides the case of two or more discrete layers each of which is optically homogeneous, we will also consider the case where the refraction index depends continuously or even smoothly on the position. This applies to both the flat and of the round earth. For the flat earth atmosphere we again use the cartesian coordinates x (horizontal) and y (vertical), assuming that the refraction index only depends on y, so $n = n(y)$. In the round case we use polar coordinates φ (horizontal) and r (vertical), assuming that $n = n(r)$.

For an optical medium consisting of a finite number of homogeneous parallel layers, the layers in the case of a flat atmosphere are separated by horizontal lines and in the round case by concentric circles, compare Figures 2.6 (page 28) and 2.7 (page 33). In both cases we will derive a *conserved quantity*, based on (2.4), $n_1 \sin\alpha = n_2 \sin\alpha'$. This quantity will be preserved when the light ray passes through the medium. As announced in Chapter 1 these conservation laws will enable us to compute the form of the light ray from the refraction index profile.

2.3.1 Atmosphere of the flat earth

For the flat earth atmosphere the cartesian coordinates (x, y) are very natural, where x is horizontal and y vertical. Our assumption then is that the refraction index n only depends on the height y, so that $n = n(y)$.

Discrete case with several horizontal layers. We consider an optical medium as in Figure 2.6, consisting of a finite number of homogeneous,

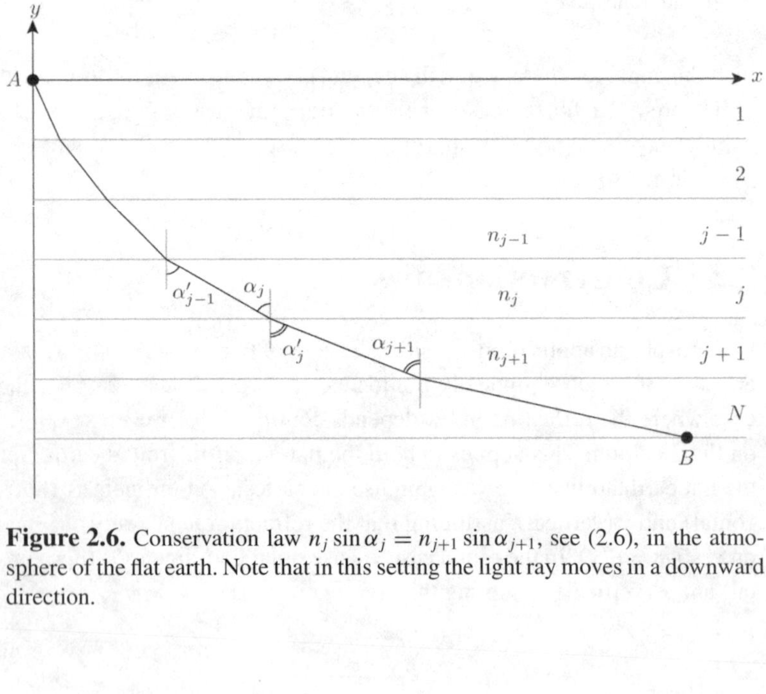

Figure 2.6. Conservation law $n_j \sin\alpha_j = n_{j+1} \sin\alpha_{j+1}$, see (2.6), in the atmosphere of the flat earth. Note that in this setting the light ray moves in a downward direction.

horizontal layers, numbered $j = 1, 2, \dots, N$. The points A and B lie in the layers 1 and N and we study the light ray from A to B as this has to run according to the weak Fermat Principle. For this we use Theorem 2.1 in the reformulation (2.4). As before in each layer the light ray forms a straight line that is traversed with constant velocity. Let n_j be the refraction index in layer j. The total light ray then is a broken straight line, where the entrance and exit angles at the boundary between layer j and $j + 1$ always will be indicated by α_j and α'_j respectively. On these boundaries, according to (2.4) we have

$$n_j \sin \alpha_j = n_{j+1} \sin \alpha'_j , \quad \text{for} \quad j = 1, 2, \dots, N - 1. \tag{2.5}$$

Since the horizontal lines are parallel, a well-known theorem of Euclidean geometry tells us that $\alpha'_j = \alpha_{j+1}$, for $j = 1, 2, \dots, N - 1$. So we obtain the following result.

Corollary 2.2 (Conservation law in horizontal layers). *Assume the above circumstances, where the points A and B lie, respectively, in the first and the Nth horizontal layer. A broken straight line from A to B is a light ray according to the weak Fermat Principle if and only if*

$$n_j \sin \alpha_j = n_{j+1} \sin \alpha_{j+1}, \tag{2.6}$$

for $j = 1, 2, \dots, N - 1$.

Remarks.

- In Figure 2.6 we only indicated the case of refraction. In case of reflection (2.5) is still valid: in that case for α we just take a supplement $\pi - \alpha$, as noted earlier.
- In case for a certain j one has $\alpha_j = \arcsin(n_{j+1}/n_j)$ where $n_j > n_{j+1}$, a limit angle occurs and the outgoing light ray is horizontal, tangent to the boundary line. Also in that case (2.6) remains valid: in fact, then one has $\sin \alpha_{j+1} = 1$.
- The remark on the extremum being a minimum as stated after Theorem 2.1 also holds in this setting.

Inclination and conservation law (2.6). In the sequence of angles

$$\alpha_1, \alpha_2, \alpha_3, \ldots \tag{2.7}$$

we may consider α_j as a parameter along the light ray. This parameter has constant values on the straight segments within a homogeneous layer and has discontinuities at the boundaries. Then (2.6) is understood as a conservation law, expressing namely that the quantity

$$S = n(y) \sin \alpha \tag{2.8}$$

is constant along the entire light ray. In the sequel the angle α is called the *inclination*, to be considered as a function of the point on the light ray. Note that in the case of refractions, where the light ray goes down we can write $\alpha = \alpha(y)$, but when reflections occur this function would become multivalued.

Two remarks are now in order. Firstly, given the refraction index profile $n = n(y)$, the *form* of the light ray is completely determined as soon as the constant value $S = n(y) \sin \alpha$ is fixed. In this way variation of S gives all possible light rays. Secondly, we recall that the velocity v of the light ray always is given by $v(y) = 1/n(y)$.

The continuous limit. Now suppose that the refraction index profile $n = n(y)$ is continuous, or even smooth. In the above we just take the limit where $N \to \infty$ and where the thickness of all layers tends to 0. The key point is that the conservation law (2.8) remains fully valid. Similar infinitesimal arguments in the seventeenth century were used by Newton, Leibniz, Bernoulli and others.

Thus the inclination α becomes a continuous (or smooth) parameter along the light ray. In the calculations, we can often eliminate it.

Remarks.

- In many cases the inclination α itself also can serve to parametrize a light ray. In Chapter 4 we shall see how this idea is explored in the brachistochrone problem, where the light ray turns out to be a cycloid.

– When α is considered as a function of y, in the case of reflection we meet with two-valuedness of $y \mapsto \sin\alpha(y)$. Below we let this happen when parametrizing the light ray as a function of y.[3]

Form of the light ray. From the conservation law (2.8) we now derive the form of the light ray as a graph $x = x(y)$. For a given value S we have

$$\sin\alpha = \frac{S}{n(y)},$$

while it is also easy to see that

$$\frac{dx}{dy} = -\tan\alpha \tag{2.9}$$

where the minus sign is due to the fact that the ray points downward. Combination of these formulas allows us to eliminate α to get

$$\frac{dx}{dy} = -\frac{S}{\sqrt{n^2(y) - S^2}}.$$

The conclusion now follows.

Theorem 2.3 (Form of a light ray: flat earth case). *In the atmosphere of the flat earth with cartesian coordinates x and y as before and with a translation invariant and continuous refraction index profile $n = n(y)$, the form of the light rays is given by*

$$x(y) = S\int_y^{y_0} \frac{d\eta}{\sqrt{n^2(\eta) - S^2}}, \tag{2.10}$$

where S is a real parameter.

In the integral (2.10) the upper limt y_0 is chosen such that $x(y_0) = 0$. We recall that in Figure 2.6 we have $y < y_0$, which corresponds to the minus sign in the equation.

[3] Think of the curve $x^2 = y$, which is the graph of a two-valued function $x = \pm\sqrt{y}$.

Remarks.

- The minus sign in (2.9) is due to the fact that the ray points downward in the setting of Figure 2.6. In the sequel situations come up where the ray points upward, for instance after the downward ray has reflected in a horizontal line. In general we have

$$\frac{dx}{dy} = \pm \tan \alpha$$

  whenever the ray describes a monotonic piece of the graph of $x = x(y)$.

- Note that the integrand of (2.10) has a singularity that occurs as soon as $n^2(y) = S^2$, in which case $\alpha = \pm\frac{1}{2}\pi$: the light ray then is perpendicular to the vertical direction which corresponds to a limit angle. In the smooth case, this usually corresponds to reflection and therefore to two-valuedness of the function $y \mapsto x(y)$. Later on we shall come back to this.

- The expression (2.10) also holds for all discrete cases. In general the function $n = n(y)$ only needs to be Lebesgue integrable.

2.3.2 Atmosphere of the round earth

In the case of the round earth, polar coordinates (φ, r), as discussed above, are suitable for describing the course of the light rays. The vertical coordinate r measures the distance to the center of the earth, while the horizontal φ–coordinate is the corresponding polar angle. As before we now assume that the refraction index n only depends on the vertical height r and the inclination α therefore is defined as the angle between the light ray and the radial direction.

Discrete case with concentric layers and the continuous limit. Again we consider an optical medium consisting of a number of homogeneous, concentric layers, numbered $j = 1, 2, \ldots, N$, where n_j is the refraction index of layer j. Compare Figure 2.7. And again the light ray is a broken straight line, where the angles with the "vertical" at the boundary between layer j and $j+1$ are indicated by α_j, respectively α'_j.

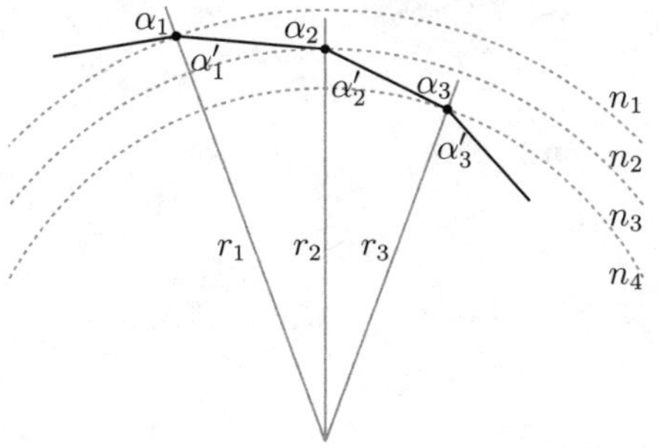

Figure 2.7. Conservation law $rn(r)\sin\alpha = C$ in the rotationally symmetric atmosphere of the round earth, see (2.14) .

The starting point again is (2.4) as this follows from Theorem 2.1 and (2.6):

$$n_j \sin\alpha_j = n_{j+1}\sin\alpha_j'. \tag{2.11}$$

For this occasion we apply the *law of sines*, another known theorem of planar geometry. As an example consider the triangle bounded by r_1 and r_2, in which we have

$$\frac{\sin\alpha_1'}{r_2} = \frac{\sin(\pi - \alpha_2)}{r_1},$$

and, more generally,

$$\frac{\sin\alpha_j'}{r_{j+1}} = \frac{\sin(\pi - \alpha_{j+1})}{r_j}.$$

Since $\sin(\pi - \alpha_{n+1}) = \sin\alpha_{n+1}$ it follows that

$$r_j \sin\alpha_j' = r_{j+1}\sin\alpha_{j+1}, \tag{2.12}$$

where r_j is the radius of the jth boundary circle, for $j = 1, 2, \ldots, N-1$. From (2.11) and (2.12) we now conclude, in analogy to Corollary 2.2, the following result.

Corollary 2.4 (Conservation law in circular layers). *Assume the above circumstances where the points A and B lie, respectively, in the first and* *the Nth circular layer. A broken straight line from A to B is a light ray according to the weak Fermat Principle if and only if*

$$n_j r_j \sin\alpha_j = n_{j+1} r_{j+1} \sin\alpha_{j+1}, \tag{2.13}$$

for $j = 1, 2, \ldots, N-1$.

Also in this case (2.13) has the format of a conservation law, namely of the quantity

$$C = rn(r)\sin\alpha\,, \tag{2.14}$$

which is constant along the entire light ray.

Many of the remarks following Corollary 2.2 also apply here.

Remarks.

- In case of the round earth the following in principle may occur. Assume that the point A is fixed and that B moves continuously away from A, starting from the case where $B = A$. Initially the extremum of the time needed to travel from A to B according to the weak Fermat Principle also is a minimum. If B moves around the earth, at some moment the situation occurs that the extremum is no longer minimal: a shorter time can be reached by letting the path from A go around the other side of the earth.[4] As we saw before, locally both versions of the Fermat Principle concur, while we see now that at larger range differences occur. We will come back to this later on. Compare this with the setting of Corollary 2.2, where this difference does not occur.
- In the form (2.14) the quantity $C = rn(r)\sin\alpha$ is conserved, also in the continuous limit of a continuous (or smooth) profile $n = n(r)$, where $N \to \infty$ and the thickness of the layers tends to 0. And as before the value of C determines the form of the light ray, which then is traversed with a velocity $v(r) = 1/n(r)$.

[4]Mathematically this is true, but in reality the light rays will be extinguished in the atmosphere after a short distance.

The form of the light ray. The story in § 2.3.1 can be repeated almost verbatim, and now leads to the light ray as the graph of a function $\varphi = \varphi(r)$, which is two-valued in the case of reflection, compare Figure 2.2, left. Here we use that

$$\sin\alpha = \frac{C}{rn(r)}, \quad \text{while in this case} \quad \tan\alpha = \pm\frac{rd\varphi}{dr},$$

as we see from Figure 2.8. This yields the following result.

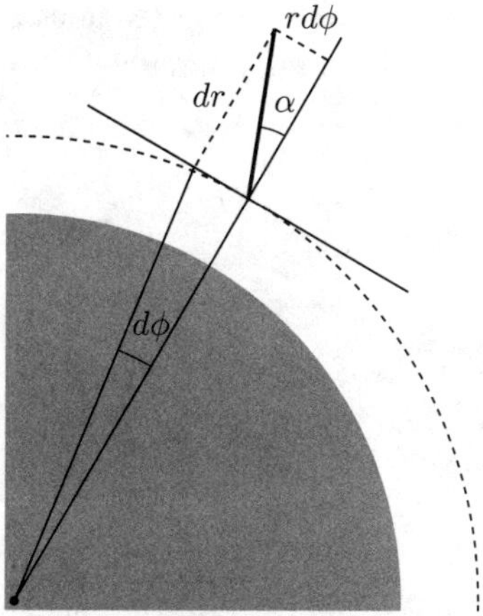

Figure 2.8. The formula $\tan\alpha = r\,d\varphi/dr$ put on the screen.

Theorem 2.5 (Form of a light ray: round earth case). *In the atmosphere of the round earth with polar coordinates φ and r as before, and with a rotation invariant and continuous refraction index profile $n = n(r)$, the form of a light ray is given by*

$$\varphi(r) = \pm C\int_{r_0}^{r} \frac{d\varrho}{\varrho\sqrt{\varrho^2 n^2(\varrho) - C^2}}, \tag{2.15}$$

where C is a real parameter.

Again we put $\varphi(r_0) = 0$. And also note that a limit angle appears at a singularity of the integrand which occurs for $r^2n^2(r) = C^2$, i.e., when

$\alpha = \pm\frac{1}{2}\pi$, which means that the light ray is perpendicular to the vertical direction: in this case the radial direction.

Remarks.

- The remarks for the case of the flat earth also apply here *mutatis mutandis*. See Figure 3.18 for an example; also see Chapter 3 and Chapter 7.
- The limit angle behavior of the light rays differs in the discrete and the continuous case. While a light ray in the discrete case by a tangency disappears into space, in the continuous case it spirals around the earth, tending towards a limiting circle. Compare Figure 2.9 and also Figure 7.3 in Chapter 7.

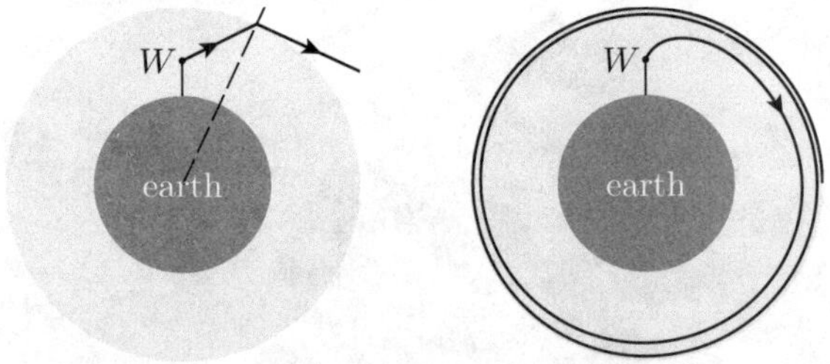

Figure 2.9. Limit angle behavior in the discrete and the continuous case. As in Figure 2.3 also now the ray comes from below. The reason for this will be explained in Chapter 3.

2.4 Scholium

The conserved quantities (2.8) and (2.14) in the above led to (2.10) and (2.15)

$$x(y) = \pm S \int_{y_0}^{y} \frac{d\eta}{\sqrt{n^2(\eta) - S^2}} \quad \text{and} \quad \varphi(r) = \pm C \int_{r_0}^{r} \frac{d\varrho}{\varrho\sqrt{\varrho^2 n^2(\varrho) - C^2}}$$

for light rays in the flat earth, respectively, the round earth atmosphere. In principle these already suffice to describe the ray bundles for a given refraction index profile. That integration in terms of explicit "elementary" functions is not always possible does not have to hold us back from applying numerical and graphical resources.

In Chapter 4 we shall see how such a computation runs successfully in the case of the brachistochrone problem. This will show that Bernoulli's own geometric-optical solution completely fits within our theory concerning the flat earth atmosphere. The translation symmetric refraction index profile then is given by energy conservation for the motion under constant gravitation and has the form

$$n(y) = \frac{1}{\sqrt{-2gy}},$$

where g is the gravitational acceleration. Also in the next example the integration runs successfully.

Example 2.1 (The Poincaré upper halfplane). In the upper halfplane $y > 0$ we consider the refraction index profile

$$n(y) = \frac{1}{y},$$

which again fits within the translation invariant theory of the flat earth atmosphere. Formula (2.10) then gives the following expression for a light ray:

$$\begin{aligned} x(y) &= \int_{y_0}^{y} \frac{d\eta}{\sqrt{\frac{1}{\eta^2} - 1}} \overset{\eta>0}{=} \int_{y_0}^{y} \frac{\eta\, d\eta}{\sqrt{1-\eta^2}} \\ &= -\frac{1}{2} \int_{y_0}^{y} \frac{d(1-\eta^2)}{(1-\eta^2)^{\frac{1}{2}}} \\ &= -\sqrt{1-\eta^2}\Big|_{y_0}^{y} \\ &= \sqrt{1-y_0^2} - \sqrt{1-y^2}, \end{aligned}$$

where

$$x_0 = \sqrt{1-y_0^2}$$

where we took $S = 1$. In this way we obtain that this light ray is a half circle

$$(x - x_0)^2 + y^2 = 1$$

that in both ends is perpendicular to the x–axis. In the general case where $S \neq 0$ we so obtain

$$(x - x_0)^2 + y^2 = \frac{1}{S^2},$$

while the case $S = 0$ leads to vertical straight line light rays.

Observant readers may recognize the geodesics of the hyperbolic metric on the Poincaré upper halfplane, see Figure 2.10. This example suggests an explicit connection between geometric optics and Riemannian geometry, For general theory see Chapter 5. Historically the Poincaré upper halfplane was one of the first examples of a non-Euclidean geometry.

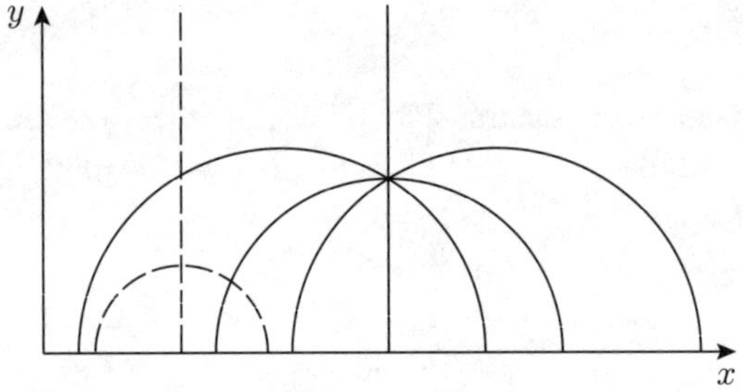

Figure 2.10. Geodesics on the Poincaré upper halfplane.

In Chapter 5 the present approach will be developed in more depth in terms of variational calculus and differential geometry, where also the velocities (i.e., the time derivatives) $\dot{x}$, $\dot{y}$, $\dot{\varphi}$ and $\dot{r}$ play a role. Here the conservation laws directly follow from translation and rotation symmetry via the Noether Principle [7]. It turns out that S and C are related to the conservation of both energy and momentum, respectively, angular momentum. The geometric character of this approach will be reinforced in Chapter 7, where the geometric optics of the round earth atmosphere will be compared with the geodesic problem on a surface of revolution.

As a bonus that sheds even more insight in the possible light ray bundles, we get a so-called *reduced phase portrait* at our disposal, which lives in the $(y, \dot{y})$–, respectively, the $(r, \dot{r})$–plane. The refraction index profile there immediately leads here to a helpful tool known as the *effective potential* [7, 43].

Chapter 3

Atmospheric optics

In this chapter we will give a deeper treatment of the atmospheric phenomena around the horizon, as discussed in Chapter 1. There we encountered blank strips or zones in the setting sun and also optical illusions like the fata morgana. Our aim is to obtain a better understanding of these phenomena, armed with the geometric optics of Chapter 2. In particular, we shall apply the mathematics around Theorem 2.5 in order to find the ray bundles associated to a given refraction index profile. Here we restrict to the case of the round earth atmosphere.

As indicated in Chapter 1, this mainly boils down to a qualitative form of modelling, where the search is for a refraction index profile that can explain the phenomena. As mentioned, it is virtually impossible to gain knowledge of this profile over the full depth of the atmosphere by observation . In this respect we spoke of an *inverse problem*.

We start with an atmosphere consisting of two concentric, homogeneous layers. Here we can already define the *Wegener-sector*, which is fundamental in this treatment. This simple setting provides qualitative explanations of most of the near-horizon phenomena.

After this we study an atmosphere consisting of several concentric, homogeneous layers. In particular a number of quasi-realistic refraction index profiles come up for discussion that go back to Takens, cf. [19]. In the latter case the atmosphere is divided into 8 or 9 layers and these examples can be viewed as approximations of cases with a continuous refraction index profile. Finally we consider the case of a continuous, even smooth, refraction index profile and present a comprehensive theory. In Chapters 7 and 8 we return to this from a deeper mathematical background. Also compare [28].

We recall from Chapter 1 that all pictures have been deliberately drawn far out of scale, in order to make the mathematical aspects clearly visible.

3.1 Fundamental concepts and terminology

We continue our symmetry assumption, which amounts to the independence of the refraction index profile of the horizontal, i.e., angular, position. As a first approximation we consider the discrete atmosphere consisting of two concentric, homogeneous layers, compare Figure 3.1.

Wegener sector. A key notion is formed by the *Wegener sector*, which consists of light rays which the observer perceives by reflection in the layer boundary above him, i.e., the lower boundary of the warm layer. Here we speak of *upper reflection*.

Sight ray. Since the only light rays that interest us are the ones that reach the observer's eye, we use a mathematical device that builds in this restriction. The key idea is that mathematically the conditions for a light ray from the eye to an object are equivalent to the reverse, i.e., the physical light rays from the object to the eye. We call these *sight rays*.

Remarks.

- Note that in the mathematics of geometric optics a sight ray also is a light ray, although physically this makes no sense: the photons only travel in one direction, towards the eye.
- In this terminology the Wegener sector consists of all sight rays that do not directly leave the atmosphere, but that are reflected first in such a layer boundary. For a detailed description see the next section.

Observation and ejection directions. The sight ray leaving the eye of the observer W, who usually resides on a height of about 10 to

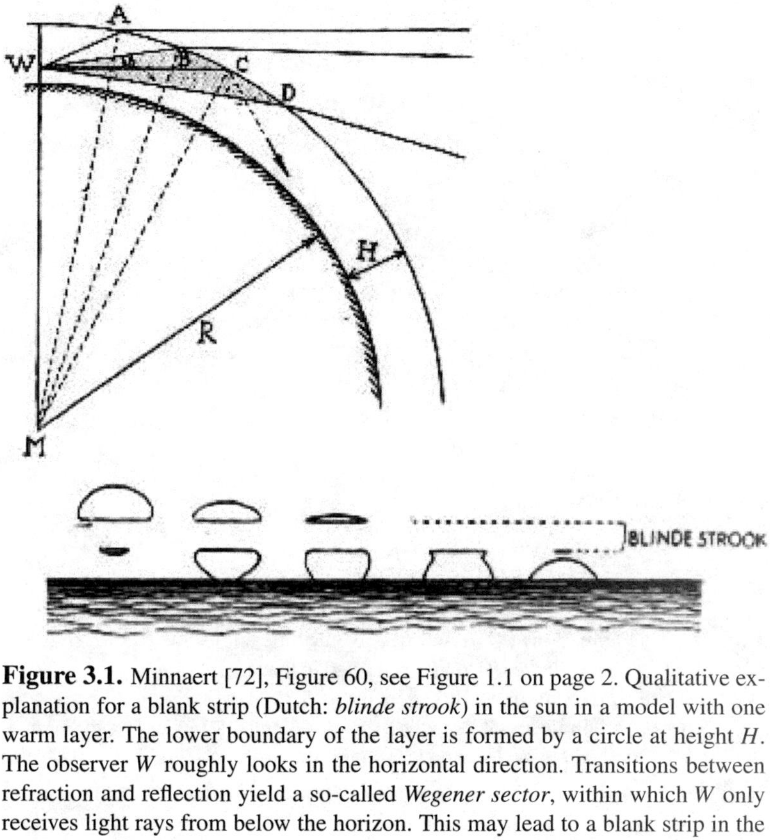

Figure 3.1. Minnaert [72], Figure 60, see Figure 1.1 on page 2. Qualitative explanation for a blank strip (Dutch: *blinde strook*) in the sun in a model with one warm layer. The lower boundary of the layer is formed by a circle at height H. The observer W roughly looks in the horizontal direction. Transitions between refraction and reflection yield a so-called *Wegener sector*, within which W only receives light rays from below the horizon. This may lead to a blank strip in the setting sun. See Chapter 1.

15 m[1], has an inclination β_W, which usually has values close to the horizontal direction $\frac{1}{2}\pi$. For this reason it is convenient to introduce

$$\delta = \frac{1}{2}\pi - \beta_W \,, \tag{3.1}$$

to be called the *observation direction*. Besides this we also need to know the inclination α of the sight ray exactly when it reaches the boundary of

[1] Say, on the deck of a ship or on a dune.

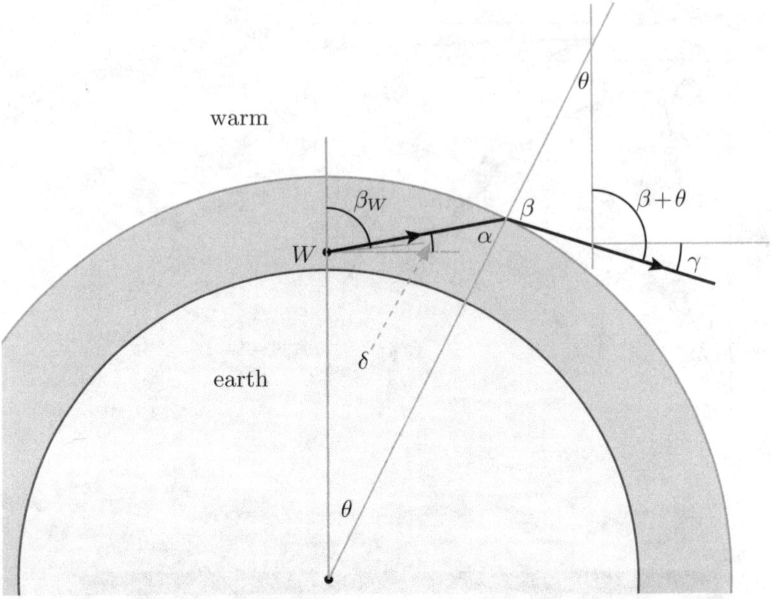

Figure 3.2. Inclinations β_W, α and β as well as the angle $\beta + \theta$ for a sight ray from the eye of the observer W, that is refracted in the lower boundary of the warm layer. Also the observation direction $\delta = \frac{1}{2}\pi - \beta_W$ and the ejection direction $\gamma = \frac{1}{2}\pi - (\beta + \theta)$ are indicated. Compare Figures 2.3 and 2.9 in Chapter 2. We note that both δ and γ vary over (usually very small) neighborhoods of 0. For examples see below.

the warm layer, see Figure 3.2 (page 44). In this boundary the sight ray is either refracted or reflected, and the inclination changes to β.

In the case of refraction, the sight ray is ejected into space, almost following a straight line. We also consider the angle $\beta + \theta$, where θ is the angle between the radius directed to W and the one directing to the point where the sight ray reaches the warm layer. As a rule also $\beta + \theta$ varies around $\frac{1}{2}\pi$, for which reason we introduce the difference

$$\gamma = \frac{1}{2}\pi - (\beta + \theta), \tag{3.2}$$

called the *ejection direction* of the sight ray.

Transition diagram. A planar figure, in which for a given refraction index profile $n = n(r)$ the value of the ejection direction γ is plotted against that of the observation direction δ, is called *transition diagram*. This diagram expresses how the image that enters through the warm layer reaches the observer W. Below we shall see how this works both for sunsets and for fata morganas.

The geometric optical theory of Chapter 2, in particular the formula (2.15) from Theorem 2.5, allows us to transform any given refraction index profile into a transition diagram together with the associated ray bundle. However, we note that for the numerical computations it is very convenient to also use the more advanced theory from Chapter 5.

3.2 Discrete description of the atmosphere

We recall that Wegener's setup has two discrete layers [93, 94], see Figure 3.4 (page 47) and also compare Figure 3.1. If M is the center of the earth, then the surface of the earth in the vertical plane of observation gives rise to a circle around M of radius R. The circular lower boundary of the warm layer has height H and thus forms a circle around M of radius $R+H$. If n_1 is the refraction index of the lower layer and n_2 that of the warm upper one, then $n_1 > n_2$: a discontinuity appears. See Figure 3.4, bottom left. This means that the setting exactly is that of the Figures 2.3 and 2.9 in Chapter 2.

Critical height. We consider all sight rays that at the lower boundary circle of the warm layer (where $r = R+H$) have an incoming limit angle inclination

$$\alpha_g = \arcsin\left(\frac{n_2}{n_1}\right), \tag{3.3}$$

according to §§ 2.2.3 and 2.2.2; compare Figure 3.3. In that case all exit sight rays are tangent to this boundary circle. If we draw all these lines, two line bundles come into being, that both envelop a circle; a circle that by definition is tangent to all these rays. The corresponding height k is called *critical height* and the corresponding circle has center M and radius

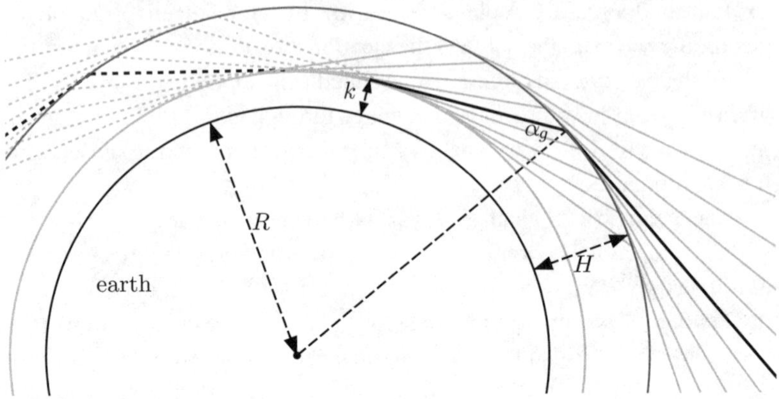

Figure 3.3. Critical circle with radius $R+k$ as envelope of the light rays with limit angle inclination α_g. The rays are divided into two line bundles, one pointing to the left (dashed) and the other to the right.

$R+k$. By construction $k < H$. The observer W moreover has a certain positive height with respect to the earth's surface, to be indicated by ε. In our main theorem below, we will need to assume that $k < \varepsilon < H$.

3.2.1 The Wegener sector

Now we can give a detailed qualitative description of the Wegener sector as depicted in Figure 3.4, top. The observer looks to the right (say, to the west) and lets his glance wander from above downwards. The sight ray thus varies from WA to WE, with A and E on the lower boundary of the warm layer. As he does this, the exit inclination β_W increases. When at the boundary point B the limit angle $\alpha = \alpha_g$ is reached, refraction of the sight ray changes into reflection in the boundary circle. In Figure 3.4 reflection still occurs when the observer W, along the ray WC, looks exactly in the horizontal direction. When we come to the ray WD the limit angle inclination $\alpha = \pi - \alpha_g$ is reached, where reflection again changes into refraction, see for instance the ray WE.

The Wegener sector $\angle BWD$ corresponds to the interval $\alpha_g < \alpha < \pi - \alpha_g$, the boundary points of which have the same sine-value. Clearly

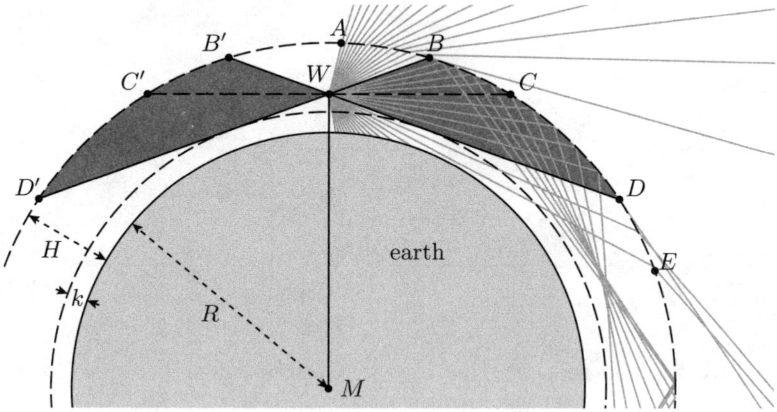

Figure 3.4. Wegener sector for $k > 0$, also compare Figure 3.1 (Minnaert [72] Figuur 60). For a qualitative description see the text.

all sight rays within the Wegener sector $\angle BWD$ reach below W's horizon. The corresponding light rays are reflected in the layer at height H (upper reflection) before they reach W. Below we shall study the optical consequences further.

The corresponding transition diagram, given in Figure 3.5 on page 48, will play an important role in this, since it describes what can be observed by W in the neighborhood of the horizontal direction $\delta = 0$. In Figure 3.5 (page 48), right, the gap in the graph exactly corresponds to the Wegener sector. We see that the corresponding interval on the δ–axis is symmetric around the value $\delta = 0$, which is in accordance with Theorem 3.1. In fact, a closer look at the triangle MWB in Figures 3.2 and 3.4 reveals that this δ–interval is given by $|\delta| \leq \frac{1}{2}(\pi - \alpha_g - \beta_W - \theta)$.

Theorem 3.1 (Discrete Wegener sector [93, 94]). *Assuming the above circumstances, a* Wegener sector *occurs for $k < \varepsilon < H$; in other words, when the observer W is in between the critical circle and the lower boundary of the warm layer. In that case we have*

$$\angle CWD = \angle CWB.$$

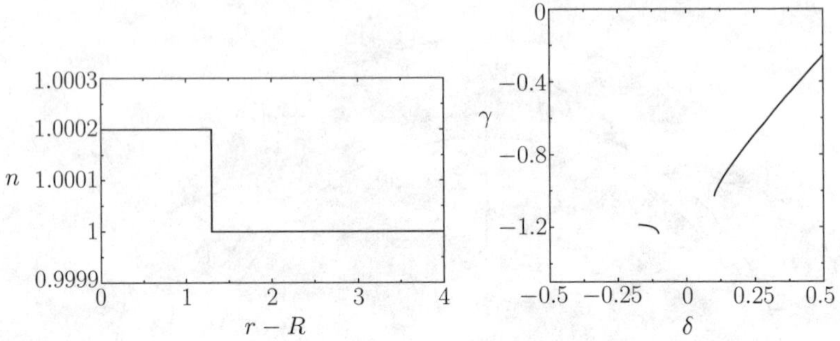

Figure 3.5. Data supporting the Wegener sector of Figure 3.4, where $k > 0$. Under this condition a blank strip can occur, as described in § 3.2.2.1. Left: Refraction index profile for $k > 0$. The unit of the horizontal axis is the hectometer. Right: Corresponding transition diagram. Along both axes we measure angles in degrees.

In other words, the Wegener sector stretches out equally far above as below the horizontal observation direction.

Proof. The straight lines WB and WD are both tangent to the critical circle. Since both pass through W we see at once that the horizontal observation direction is the bisector of $\angle BWD$. Indeed, from symmetry considerations it follows that MW is the outer bisector of $\angle BWD$, and since this is perpendicular to WC, the latter must be the bisector. □

3.2.2 Optical scenarios near the horizon

We now come to a number of possible consequences of the existence of a Wegener sector for optical phenomena near the horizon. Following Minnaert [72], we start with sunsets. After this also fata morganas will come up for discussion.

3.2.2.1 The setting sun

Let us consider the consequences of a Wegener sector for the setting sun. First of all we see that the sight rays within a Wegener sector eventually disappear in the mists of the world, which are, of course, much darker than the sun. So if the sun passes the corresponding ejection directions, W perceives a blank strip or zone. In other words, the sight rays in the Wegener sector do *not* end up in solar light, so light rays from the sun do not reach the eye of W.

The transition diagrams give a more detailed impression of how this goes. The disk of the setting sun corresponds to an interval that has a maximal length of 0.5^o, and that slides down along the γ–axis. Along the δ–axis we can see how W observes this.

Blank strip. In the above scenario for $k > 0$ we are dealing with a transition diagram of Figure 3.4 and 3.5. The interval that slides down along the γ–axis, corresponding to the setting sun, at a given moment breaks into two pieces on the δ–axis as shown in Figure 3.6; the intermediate, dark or blank, interval exactly corresponds to the Wegnener sector. Rotation of the vertical plane through W over the full breadth of the solar disk hence yields a sunset with a blank strip as depicted in Figures 1.1 and 1.2 of Chapter 1. See below for alternative scenarios based on this same principle.

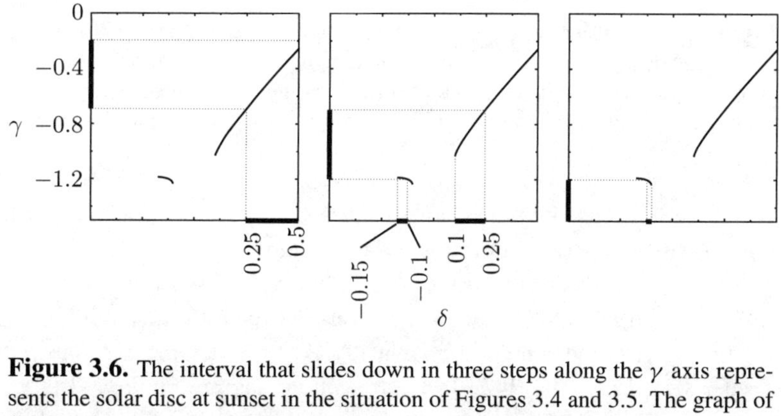

Figure 3.6. The interval that slides down in three steps along the γ axis represents the solar disc at sunset in the situation of Figures 3.4 and 3.5. The graph of the transition diagram in Figure 3.5, right, then gives rise to corresponding intervals on the δ–axis: this is what W observes of the setting sun, which in the central picture clearly has a blank strip.

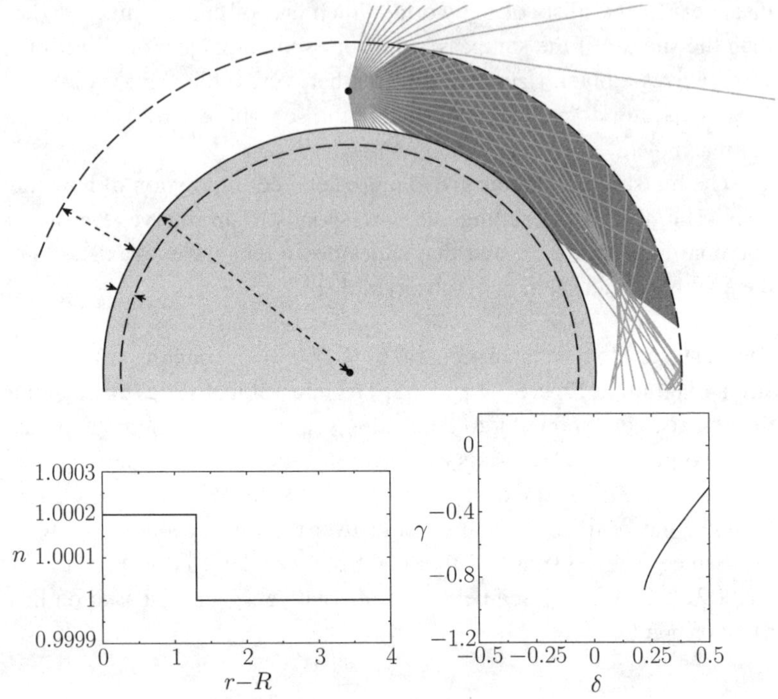

Figure 3.7. Wegener sector for $k < 0$, as a variation on Figures 3.4 and 3.5, with the same lettering and with corresponding refraction index profile and transition diagram. In this and later figures we also use the same units as before. This situation can lead to a sunset with a blank zone as depicted in Figure 1.3.

Blank zone. The situation changes when $k < 0$, i.e., when the critical circle is situated below the earth's surface, compare Figure 3.7 (page 50). The refraction index profile is the same as in Figure 3.4, but the transition diagram is different. We note that Theorem 3.1 remains fully valid.

In this case the limit angle sight ray WD reaches the earth, as do deeper seated rays like WE. In the transition diagram Figure 3.7 (bottom, right) this is expressed by the fact that, in comparison with that of Figure 3.4, the branch of the graph for negative values of δ is missing.

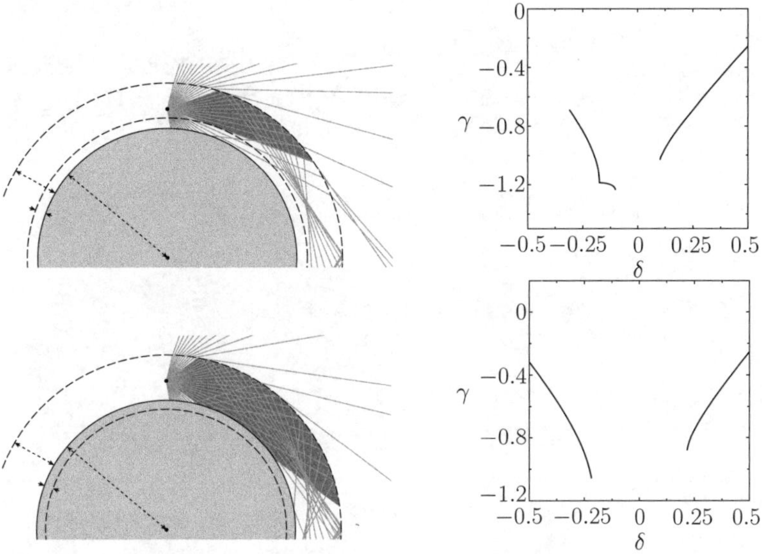

Figure 3.8. Wegener sectors for $k > 0$ and $k < 0$ as before, now including cases with lower reflection. Corresponding transition diagrams have been added; the refraction index profile always is the same as in Figures 3.5 and 3.7. The steep branches to the left correspond with lower reflection. The negative slope indicates that the image is upside down.

In the setting sun thus a *blank zone* arises. Such a scenario was already reproduced in Figure 1.3. Blank zones occur quite frequently.

Remarks.

– A closer look at the proof of Theorem 3.1 reveals that the part of the sun above the blank strip is somewhat compressed in the vertical direction, while the part below the blank strip is, on the contrary, stretched. This is reflected in the different slopes of branches in the graphs of Figures 3.4, 3.7 and 3.8 (page 51).

– If we set up the anologue of Wegener's two layers setup in the flat earth's atmosphere, we find $k = -\infty$ for the critical height. This means that on the flat earth only blank zones can occur and no strips.

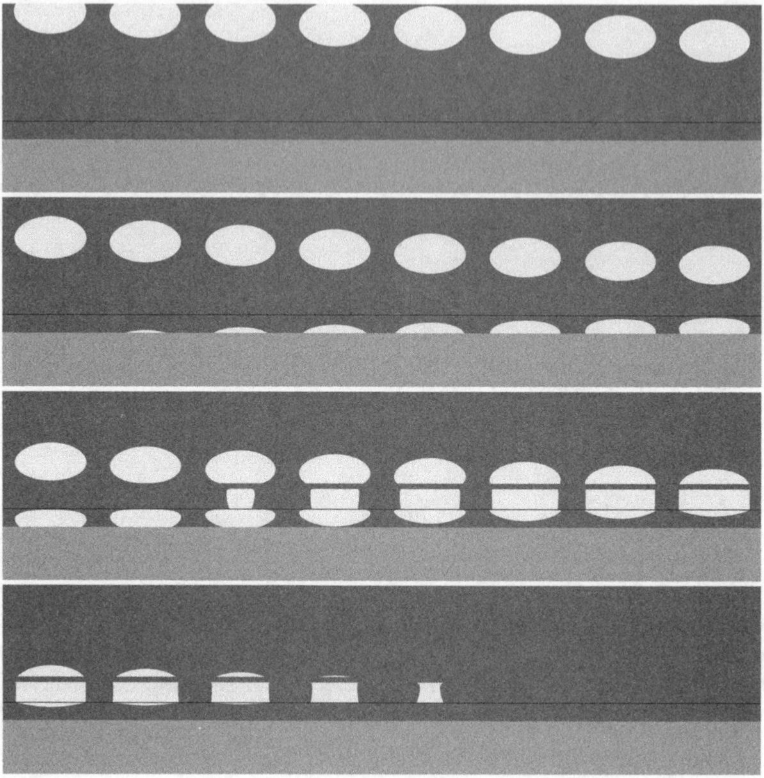

Figure 3.9. Sunset for the case of Figure 3.8, top. The bottom case of Figure 3.8 may give rise to the sunset of Figure 1.5.

3.2.2.2 Including lower reflection

Besides upper reflection in a warm layer at height H also a *lower reflection* may occur in a warm layer immediately above the earth's surface: one may think of warm desert sand or a warm sea surface. Also a smooth sheet of ice or a mirroring lake might do this. In Figure 3.8 ray bundles have been depicted for the cases $k > 0$ and $k < 0$. In each case we give a corresponding transition diagram, noting that the refraction index profile still is the same as before.

3.2.2.3 Fata morganas

By combination of refraction and reflection the observer W can sometimes see several images of objects below the horizon. We already noticed this in connection with Figure 1.5, where the upper boundary of the solar disk by lower reflection is still visible, while the sun had already set.

By a *fata morgana*[2] we generally understand an image originating below the horizon that by reflections and refractions comes to the observer. It may be of objects like a mountain range, a city, an oasis, a clump of trees, a ship, etc., upright or upside down according to whether the number of reflections is even or odd. By a combination of these factors the image may be considerably distorted. For examples of this see Figures 3.10 (page 54) and 3.11 (page 55).

Wegener sectors can play a role in this. In Figure 3.12 (page 56) we summarize four cases, presenting sight ray bundles for positive and negative k and both with and without lower reflection. In all cases the ray bundles have been made quite long. This shows that these bundles do not leave the atmosphere, something that is also easily proven. It follows that in this way optical illusions can be observed only of objects near the earth. This happens most easily in cases where $k < 0$, since then some of the sight rays hit the earth's surface.

The black sight rays of Figure 3.12 that lie just outside the Wegener sector can leave the atmosphere and some can reach into space. In a number of cases these rays can also reach below the horizon and so may give rise to celestial fata morganas, see Figure 3.13 on page 56. For instance it should be possible in this way to observe the moon while it is below the horizon. At the end of this chapter we come back to the subject of celestial illusions when describing the so-called Nova Zembla phenomenon.

[2] The nomenclature goes back to Morgan le Fay (Fata Morgana in Italian), a fairy from the King Arthur cycle. It is rumoured that she would cause optical illusions in the Strait of Messina, visible from the Sicilian coast.

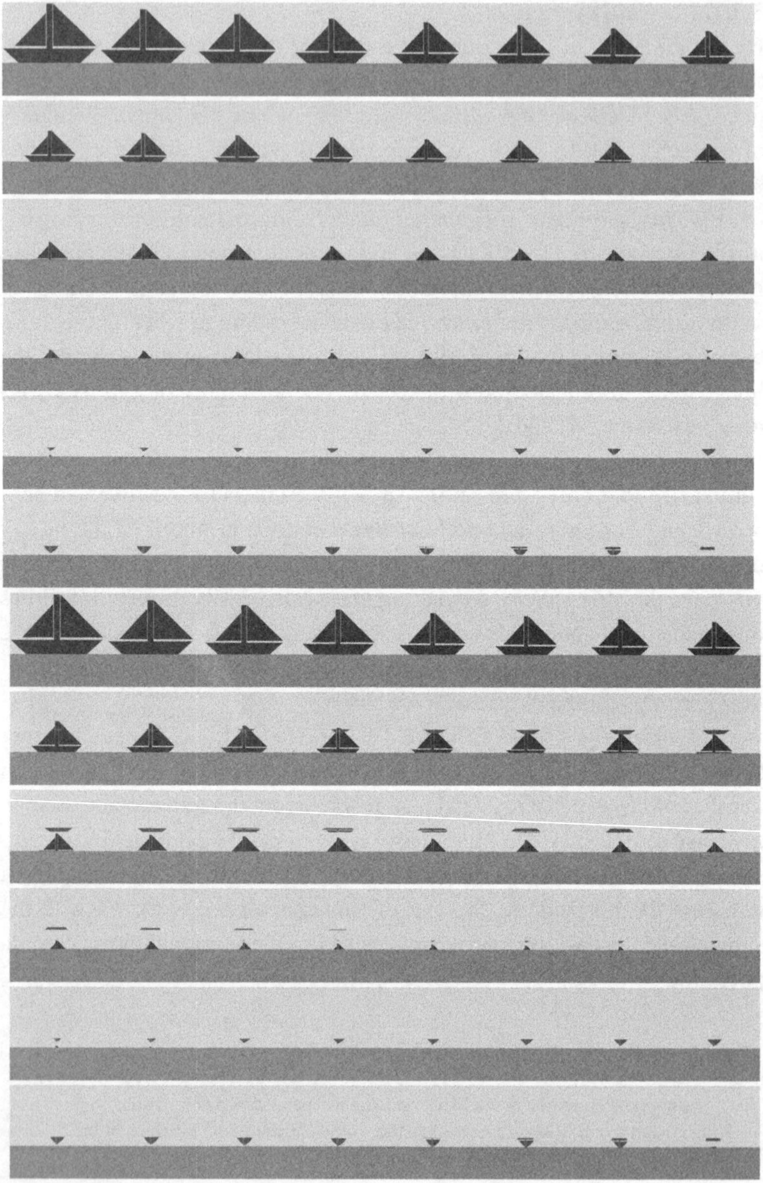

Figure 3.10. Fata morganas of a sailing ship that disappears below the horizon for $k > 0$. Top without and bottom with lower reflection.

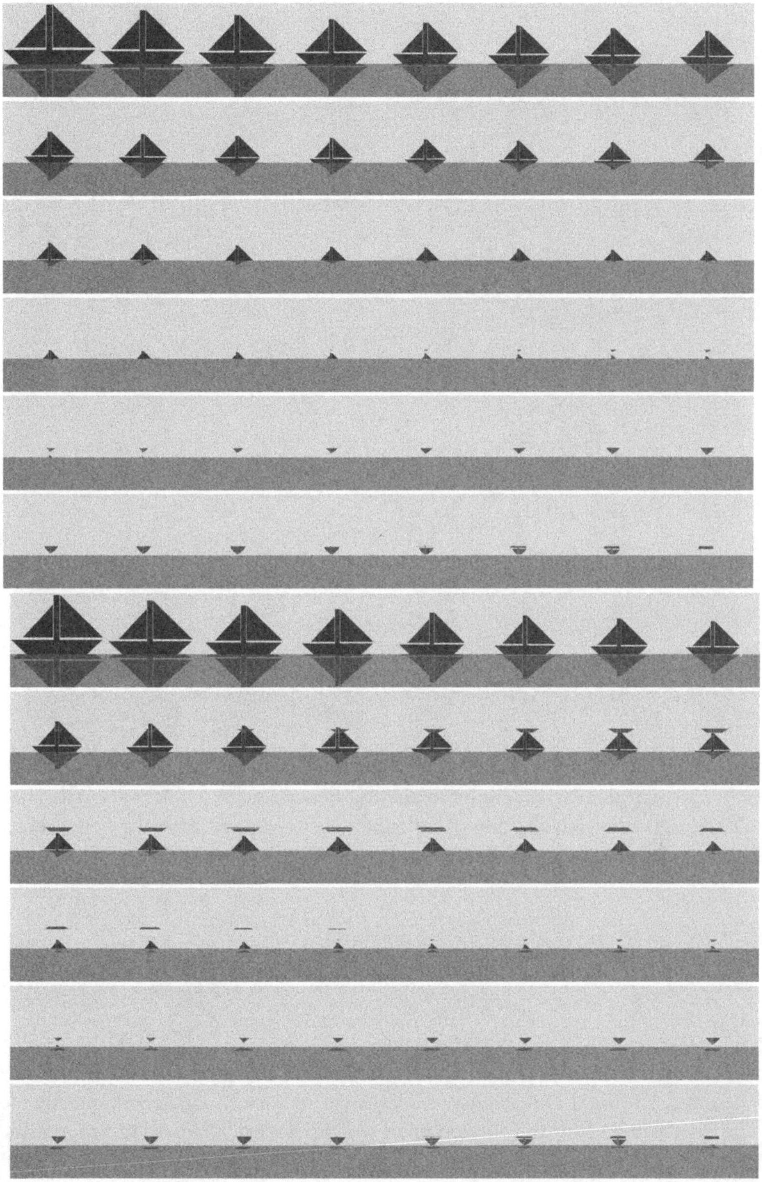

Figure 3.11. As in Figure 3.10, now with $k < 0$.

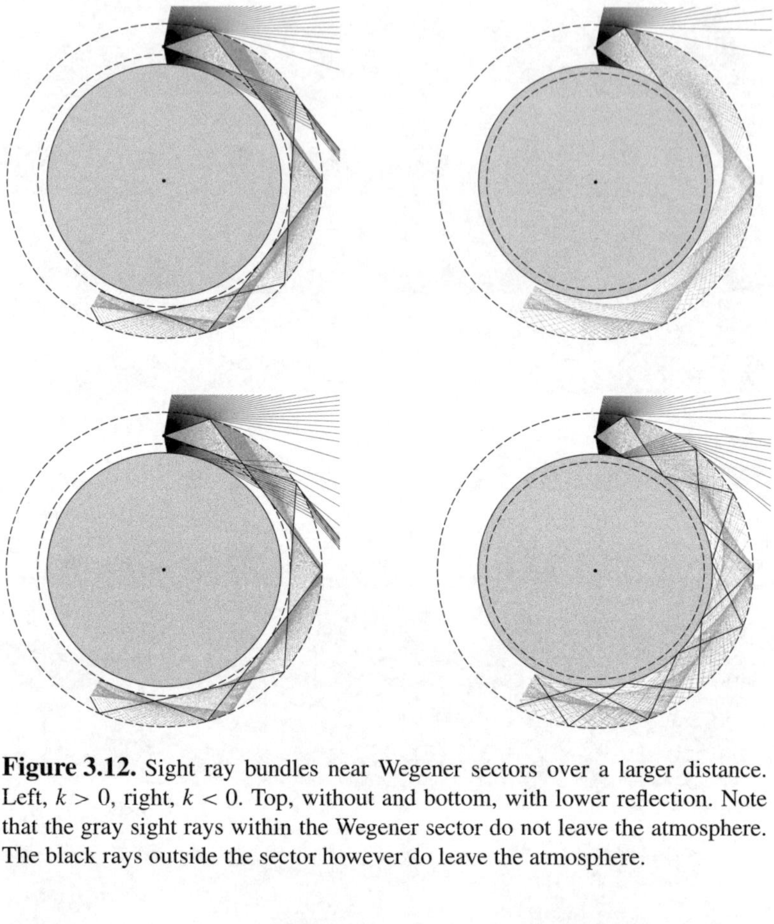

Figure 3.12. Sight ray bundles near Wegener sectors over a larger distance. Left, $k > 0$, right, $k < 0$. Top, without and bottom, with lower reflection. Note that the gray sight rays within the Wegener sector do not leave the atmosphere. The black rays outside the sector however do leave the atmosphere.

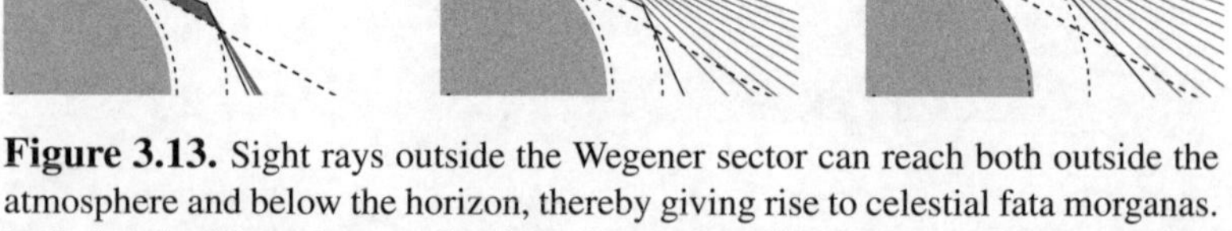

Figure 3.13. Sight rays outside the Wegener sector can reach both outside the atmosphere and below the horizon, thereby giving rise to celestial fata morganas. Left: $k > 0$ without lower reflection. Center: $k > 0$ with lower reflection. Right: $k < 0$ with lower reflection.

Table 3.1. Refraction index profiles in the cases I and II, expressed in the height ℓ above the earth.

ℓ		n
case I	case II	
0-74 m	0-73 m	1.00027
74 m-1 km	73 m-1 km	1.00026
1-3 km		1.000235
3-5 km		1.000178
5-7 km		1.000135
7-9 km		1.000105
9-11 km		1.000078
$\geq$ 11 km		1

3.2.3 Variations

Wegener's schematic approach is based on a rotationally symmetric, discrete atmosphere with two layers. From such a setup one can only expect qualitative explanations and it is surprising to see how neatly such an explanation can fit the observed phenomena. The theory of geometric optics as presented in Chapter 2, however, easily allows for refinements of Wegener's approach, particularly when the corresponding mathematics is used in numerical algorithms. Below we shall present a number of models with 8 or 9 layers that present a somewhat more realistic construction of the atmosphere. These models were developed by Floris Takens [19].

3.2.3.1 Takens–profiles

We now consider four mutually related examples, numbered I, II, III and IV, of rotationally symmetric refraction index profiles for the round earth's atmosphere. The atmosphere has been divided into a larger number of homogeneous layers and this could be considered as an approximation of a continuous profile.

The profiles numbered I and II, with 8 layers, are presented in Table 3.1, where the difference between the two is that a warm layer with

Table 3.2. The cases III and IV are like the case I and II from Table 1, only now in both cases a warm layer of 1 cm has been added just above the earth's surface.

ℓ case III	ℓ case IV	n
0-1 cm		1.000268
1 cm-74 m	1 cm-73 m	1.00027
74 m-1 km	73 m-1 km	1.00026
1-3 km		1.000235
3-5 km		1.000178
5-7 km		1.000135
7-9 km		1.000105
9-11 km		1.000078
$\geq$ 11 km		1

1 km of thickness begins at a height of either on 74 or on 73 m. Observer W is situated at a height of 10 m and perceives a light ray bundle of $32' \approx 0.5^o$ that varies around the horizontal direction as

$$\delta = -0.25^o, -0.24^o, -0.23^o, \ldots, 0.24^o, 0.25^o.$$

The refraction index profiles III and IV consist of 9 layers and are given in Table 3.2. Here profile III corresponds to I and profile IV to II, except that in both cases a warm layer of 1 cm thickness has been added immediately above the earth surface.

In Figure 3.14 on page 59, for all cases I, II, III and IV the transition diagrams are represented. Since 1 cm, or even 1 m, is so small in comparison to the kilometers in our scale, the refraction index profiles for our four cases are indistinguishable when we graph them – but the transition diagrams are not! It is striking that in the cases I and III no Wegener sector appears, while in the cases II and IV this does happen, namely in the observation directions

$$-0.03^0 \leq \delta \leq +0.03^0.$$

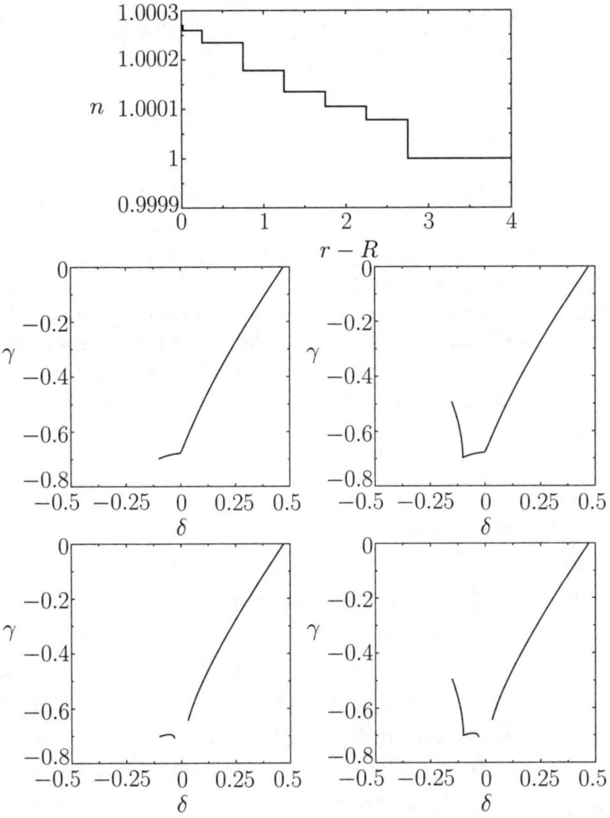

Figure 3.14. Top: Takens refraction index profile. Center: Transition diagrams for the cases I and III; bottom: idem dito for II and IV.

We see that varying the height of the warm layer by only 1 m makes a lot of difference.

The steep branches to the left of the cases III and IV again indicate lower reflection, where the negative slope means that for *W* the image is upside down, also compare Figure 3.8. The sight ray bundles associated with Figure 3.14 on page 59 are too narrow for a good presentation, but qualitatively they are completely similar to those of the earlier Figures 3.4, 3.7 and 3.8.

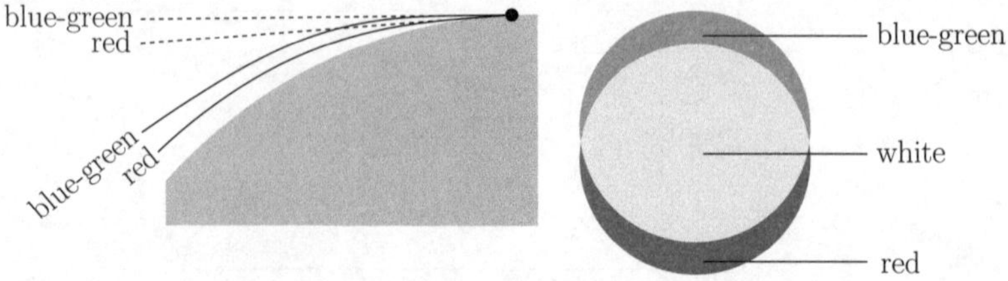

Figure 3.15. Explanation of the green flash, compare Minnaert [72], Figure 66. By the frequency dependence of the refraction index, one roughly observes two solar disks, above a mainly green one and below a red one, that largely overlap thereby giving rise to the well-known white light. The last thing we see at sunset is a small green disk. To answer the question to what extent the occurrence of a blank strip can reinforce this phenomenon, compare the various sunset scenarios described above.

3.2.3.2 The green flash

The green flash is a phenomenon of the setting (and sometimes also the rising) sun, where at the *moment suprême* a green flash appears, as briefly discussed in Chapter 1. This phenomenon is described well by Minnaert [72], § 47. One of the most important factors in the occurence of the green flash is that red light rays in the atmosphere are bent more than green ones: the refraction index (and the velocity of light) is frequency dependent. This effect gets more visible when the light rays come from the vicinity of the horizon. Roughly this gives rise to two suns, one red and the other green, see Figure 3.15 on page 60. Usually both suns coincide and thereby give the impression of a white sun: after all, red and green are complementary colors. Similar curved light rays were already discussed by Huygens [56], see Figure 4.13 in the next chapter.

In general this explains why the lower boundary of a setting sun is red, while the upper boundary has a green color. At sunset there may be a fleeting moment where the 'red' sun is already below the horizon, while the upper boundary of the 'green' one is still visible. In connection with a number of other atmospheric circumstances, one of them being an exceptionally good visibility, this can lead to a *green flash*. The existence of a Wegener sector might also play a role in this, as we shall explain below. We recommend reading also Jules Verne's *Le Rayon Vert* [92] and note that the green flash is somewhat wrapped in a veil of mysticism. Compare O'Connell [77] for a fine collection of observations from the Vatican Observatory in Castel Gandolfo, 450 m above sea level. For more information and beautiful color pictures and movies the reader is advised to search Google under the keywords 'green flash'or 'green ray'.

Relationship with a Wegener sector? The question that is in order now is whether the presence of a Wegener sector could contribute to the clearer occurrence of a green flash. The fact that the flash may well pitch upward seems to indicate a lower reflection that can occur, amongst others, in case of a blank zone, compare Figure 1.5 as this corresponds with the lower case of Figure 3.8. Also compare case IV of Figure 3.14, where the upper, green, boundary of the sun that has already set becomes visible by lower reflection.

3.3 Continuous description of the atmosphere

The continuous (usually even smooth) description of the sight rays in the round earth atmosphere runs similar to that of the discrete one; compare Figures 3.4 and 3.7. Under suitable conditions also here we will identify a Wegener sector, with the same properties as before: blank strips and zones in the solar disk, as well as earthly or celestial illusions. The present approach is fully based on the conservation law of Chapter 2, and will turn out to comprise that of § 3.2. In particular this holds for Theorem 3.1.

3.3.1 The conservation law revisited

We assume that the optical medium is isotropic and has rotational symmetry. This gives us a refraction index profile $n = n(r)$, where we recall that the speed of propagation is given by $v(r) = 1/n(r)$ (recall that the speed of light has been rescaled to $c = 1$). Then we have a conserved

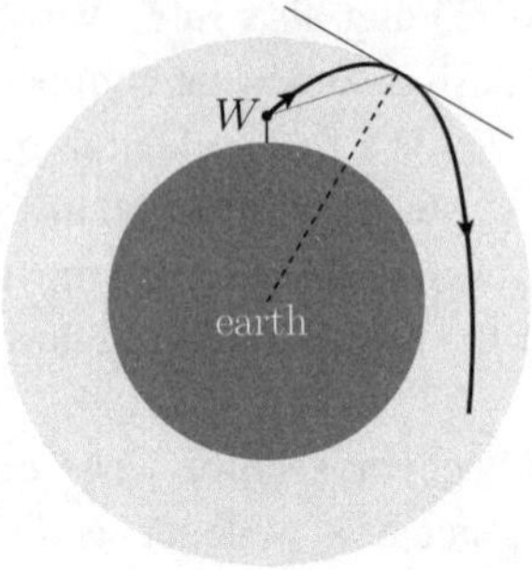

Figure 3.16. Wegener sight ray in the continuous case.

quantity (2.14)

$$C = rn(r)\sin\alpha\,.$$

Theorem 2.5 completely determines the form of the sight rays. In particular this form in polar coordinates (φ, r) is given by (2.15)

$$\varphi(r) = C\int_{r_0}^{r}\frac{d\varrho}{\varrho\sqrt{\varrho^2 n^2(\varrho) - C^2}}\,,$$

With the help of this[3] from a given refraction index profile we can determine a transition diagram, where we may assume that the outward sight rays will continue in approximately straight lines.

3.3.2 Wegener sight ray

A *Wegener sight ray* is defined as a sight ray that cannot escape directly from the atmosphere, but is reflected in a warm layer. After this the ray can land on the earth's surface or disappear into space, see Figure 3.16. In this way it may give rise to optical illusions. The union of all Wegener sight rays in the continuous case forms the Wegener sector.

The sight ray can be characterized as a parametrized curve

$$t \mapsto (\varphi(t), r(t)),$$

[3] We also use the somewhat more advanced methods of Chapter 5.

where t denotes time.[4] It now is convenient to introduce the notion of *modified refraction index profile*, defined as

$$n_M(r) = rn(r)\,. \tag{3.4}$$

In a later chapter we will see that n_M has a geometric meaning. The function $r \mapsto n_M(r)$ is a necessary ingredient for the formulation of our results. Indeed, if we define

$$C(t) = n_M(r(t)) \sin \alpha(t),$$

then the function $C(t) \equiv C(t_0)$ is constant and therefore completely determined by its value at $t = t_0$, associated with $r_0 = r(t_0)$ and $\alpha_0 = \alpha(t_0)$. Here the observer W is situated at height r_0 'emitting' a sight ray that at the instant t_0 has the inclination α_0. As before we let W look to the right[5] and thus $0 \leq \alpha_0 \leq \pi$, see Figure 3.16. Then the following result holds true.

Theorem 3.2 (Continuous Wegener sector). *Assume the above circumstances with a smooth refraction index profile $n = n(r)$. If the modified refraction index profile n_M, see* (3.4)*, has an absolute minimum in $r_m \geq r_0$, then for*

$$\alpha_g = \arcsin\left(\frac{n_M(r_m)}{n_M(r_0)}\right) \tag{3.5}$$

the interval $[\alpha_g, \pi - \alpha_g]$ corresponds to a Wegener sector. This sector is symmetric with respect to reflection in the observation direction $\alpha_0 = \frac{1}{2}\pi$.

Proof. The sight ray is non–Wegener whenever it can escape directly from the atmosphere, which means that for all $t \geq t_0$ one has that $\alpha(t) \neq \frac{1}{2}\pi$. This is equivalent to saying

$$n_M(r_0) \sin \alpha_0 \equiv n_M(r(t)) \sin \alpha(t) < n_M(r(t)),$$

[4] In Part II we will meet other parametrizations.

[5] Say, he is looking west.

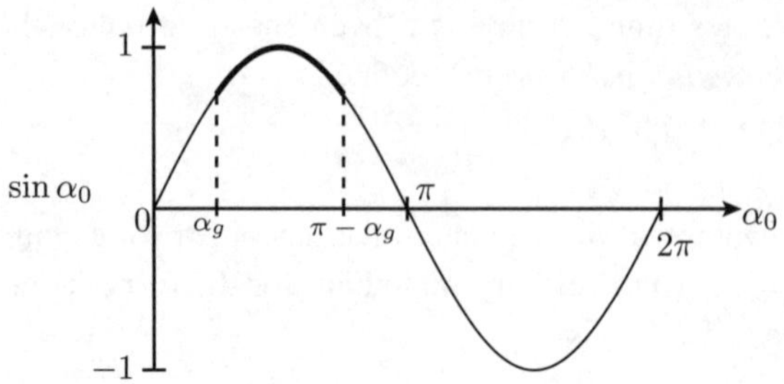

Figure 3.17. Continuous Wegener sector $\alpha_g \leq \alpha_0 \leq \pi - \alpha_g$ in terms of the graph of the function $\alpha_0 \mapsto \sin\alpha_0$.

or to

$$n_M(r_0)\sin\alpha_0 < n_M(r)$$

for any $r > r_0$. In other words, the Wegener sight rays have an initial inclination α_0, such that for some $r > r_0$

$$n_M(r) \leq n_M(r_0)\sin\alpha_0.$$

Since r_m is the absolute minimum of the function n_M it follows that

$$0 < \frac{n_M(r_m)}{n_M(r_0)} \leq 1$$

and a brief inspection of the graph of $\alpha_0 \mapsto \sin\alpha_0$, in particular the symmetry property $\sin\alpha_0 = \sin(\pi - \alpha_0)$, shows us that precisely for all α_0 with

$$\alpha_g \leq \alpha_0 \leq \pi - \alpha_g$$

a Wegener sight ray occurs, see Figure 3.17 (page 64). □

The interval $[\alpha_g, \pi - \alpha_g]$ indeed is symmetric with respect to the midpoint $\alpha_0 = \frac{1}{2}\pi$. We note that in the exceptional case where the function increases monotonically from the initial value $n_M(r_0)$ on, then $r_m = r_0$ and we have $\alpha_g = \frac{1}{2}\pi$ which reduces the interval $[\alpha_g, \pi - \alpha_g]$ to the singleton set $\{\frac{1}{2}\pi\}$ and thus the Wegener sector to the horizontal halfline. The inclination α_g as defined by (3.5), with some good will could be called the limit angle, which in the case of Theorem 3.1 exactly corresponds to $\angle CWB$.

The entire setup of Theorem 3.2, including its proof, is also applicable to Theorem 3.1, including cases with more than two discrete layers. This follows from the fact that the formulas (2.14), (2.15), and Theorem 2.5, also hold in that case.

Remarks.

- The existence of a global minimum of the function n_M surely holds in the atmospheric situation, where we have that $\lim_{r\to\infty} n(r) = 1$.
- Since outside the atmosphere we even have $n \equiv 1$, it here is also possible to define an ejection inclination and to give transition diagrams also in the continuous case. As in the discrete case this ejection inclination equals the value of $\alpha(t)$ when the ray leaves the atmosphere. Also we can easily define obervation and ejection directions δ and γ.

Example 3.1 (Continuous Wegener sector). The simplest continuous example of a Wegener sector will be given in a qualitative way for a modified refraction index profile n_M as in Figure 3.18 (page 66), bottom left. This function has exactly one maximum and gives rise to a sight ray bundle as in the top and bottom right of the same figure. This example will be discussed further in Chapter 7 of Part II.

3.4 Scholium

For further reading we again refer to Minnaert [72], part 1, Wegener [93, 94], and to Floor [39–41] and Van der Werf [95, 96].

In the past two chapters we developed a virtually complete geometric optics of the atmosphere, given the restriction to a (vertical) plane and under the extra assumption of a rotationally symmetric refraction index profile. This approach ran by the Laws of Hero and Snell, that both were based on the Fermat Principle.

Let me comment on the regularity of the refraction index profile $n = n(r)$. In principle this should be a continuous function, perhaps with

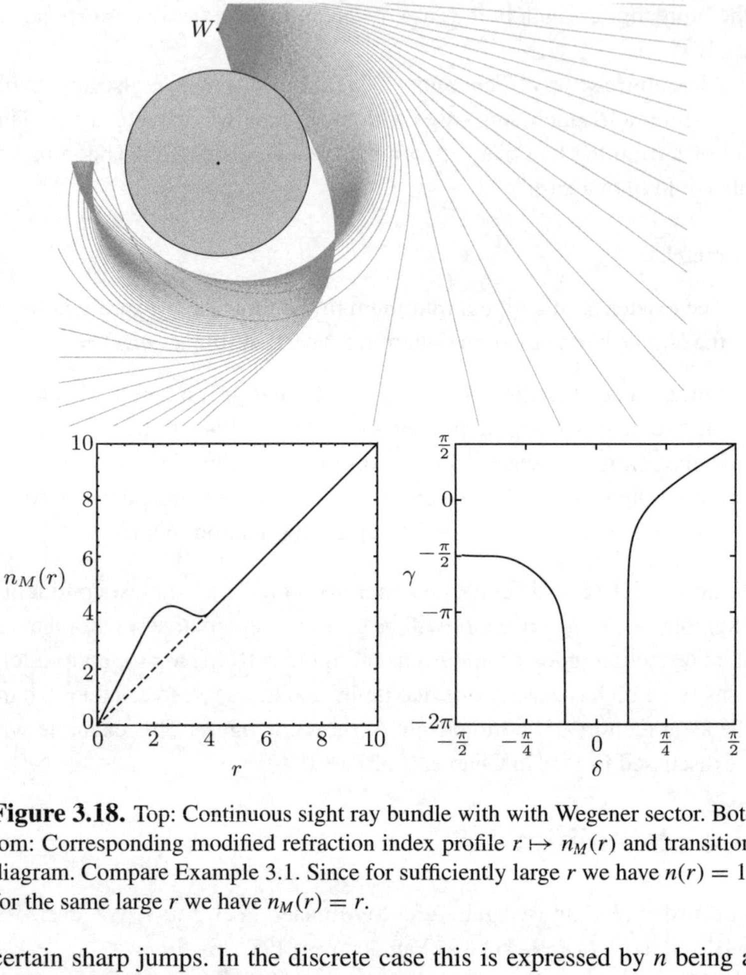

Figure 3.18. Top: Continuous sight ray bundle with with Wegener sector. Bottom: Corresponding modified refraction index profile $r \mapsto n_M(r)$ and transition diagram. Compare Example 3.1. Since for sufficiently large r we have $n(r) = 1$, for the same large r we have $n_M(r) = r$.

certain sharp jumps. In the discrete case this is expressed by n being a step function. Step functions with many steps, like the Takens profiles, can be used to approximate the continuous case. From now on the profiles will be assumed continuous, even smooth.

3.4.1 Deepening

In Part II, *Light rays as geodesics*, in particular in Chapter 5, this approach will be studied in depth with the Principle of Hamilton and some

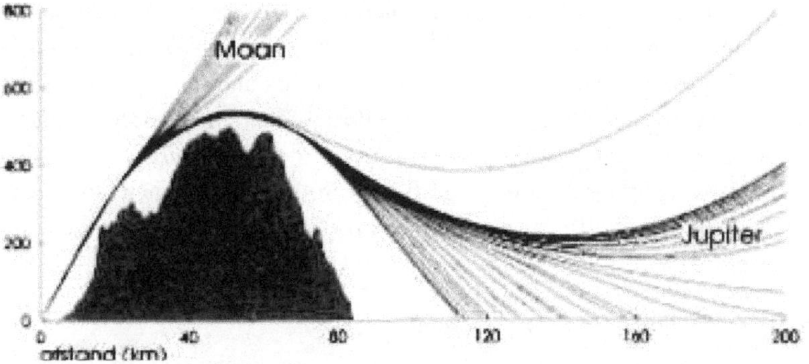

Figure 3.19. Van der Werf [95], p. 96: Sight ray bundle from an observer in the origin that can explain a conjunction of Jupiter and the moon. In this picture the earth has been rectified for convenience. The rays follow the contours of a mountain range on Nova Zembla.

differential geometry. The sight rays then become geodesics of a Riemannian metric that encodes the refraction index profile. This will enable us to drop the assumption of rotational symmetry, even that of the restriction to a vertical plane. Compare § 2.4.

In Chapter 7 we further elaborate the rotationally symmetric, planar theory, developing an analogy between sight rays and the geodesics on a surface of revolution in three-dimensional space. It is in this setting that the modified refraction index profile $n_M(r)$ gets a nice geometric interpretation that further clarifies certain aspects of the theory.

3.4.2 Nova Zembla phenomenon

The Dutch sailor Willem Barentsz ventured to sail around the north of Europe and Asia to find the Indies. In this attempt, during the winter of 1596-97 he and his crew were caught in the ice around the island of Nova Zembla and had to winter there, building a log cabin from wood taken from the ships. During the long winter months they had plenty of time to study the skies and they observed a few remarkable phenomena that were not in accordance with the astronomical tables. In particular this concerned a conjunction of Jupiter and the moon and a premature observation of the solar disk. Both phenomena, according to their tables, took place *below* the horizon at the time they were observed. For a nice historical description and good graphics of this and other celestial illusions see Van der Werf [95]. Also see [96], which intrdocuced the name *Nova Zembla phenomenon*. Note that in the mathematical reconstruction resulting in Figure 3.19 on page 67, the rotational symmetry has been abandoned; in fact the refraction index profile here depends on the landscape.

Chapter 4

Light and the cycloid

The brachistochrone problem asks for the 'fastest chute' connecting two given points in a vertical plane, see Figure 4.1 on page 70. Think of a bead K that slides along a wire profile connecting two points A and B. The sliding is assumed frictionless, so only constant vertical gravity influences the motion. The question then is what shape of wire makes the *travel time* shortest. Many people's first guess is that the answer is simply given by the straight line connecting A and B, since this corresponds to the shortest *distance*, but it turns out that the correct answer is a cycloid.

This problem was put forward in 1696-97 by Johann Bernoulli. In the present chapter we treat his own solution, which mixes classical mechanics and geometric optics in a surprising way. The interesting history of the brachistochrone problem and its considerable significance will be discussed at the end of this chapter.

We first briefly discuss the cycloid. Then we turn to geometric optics of the flat earth atmosphere. We will have to chose an appropriate refraction index profile to achieve the brachistochrone curve as a light ray. The Fermat Principle of shortest time provides the key idea.

4.1 The cycloid

To be able to understand Bernoulli's solution of the brachistochrone problem we first have to share his knowledge of the cycloid. The cycloid is the curve traced by a point on the rim of a wheel that rolls along a straight

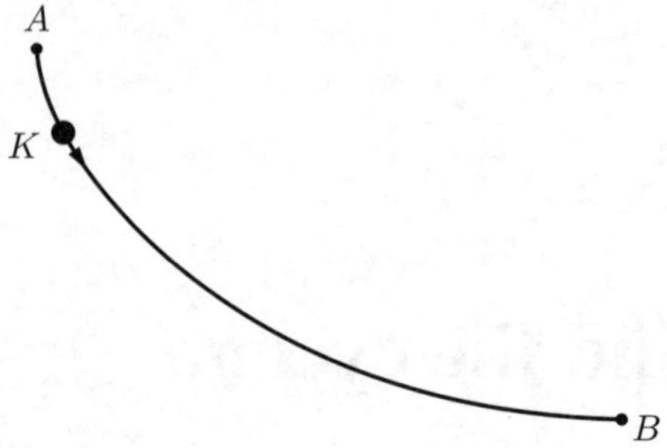

Figure 4.1. Bernoulli's brachistochrone problem [14]. Also compare Figure 1.8 on page 12.

line. Thinking of a somewhat simplified bicycle wheel, we call this point the *valve*. We let the wheel roll along the bottom of a horizontal line in a vertical plane, as sketched in Figure 4.2 on page 71. The coordinates x and y are chosen as follows: x is horizontal and y vertical, where the origin $(x, y) = (0, 0)$ marks the point where the valve on the rim is at its lowest position. If the wheel has radius ϱ, we get the parametric representation

$$x(\theta) = \varrho(\theta + \sin\theta), \; y(\theta) = \varrho(1 - \cos\theta). \tag{4.1}$$

To see this we use a bit of plane geometry. When the wheel moves along the top horizontal line to the right, the valve moves to the right as well, see the top of Figure 4.2. The line connecting the center of the wheel to the valve translates and rotates to the right, as indicated. If we now attach a wedge with opening angle θ to this line, this wedge performs the same combination of motions. Let us consider the situation when the wheel has rotated over an angle θ. First we note that the other 'leg' of the wedge then has reached the vertical position. The wedge then has translated horizontally over a distance $\varrho\theta$. Combining this with the fact that the valve meanwhile has rotated over an angle θ easily gives the above parametrization. We refer to ϱ and θ as the *radius* and the *rolling angle* of this cycloid.

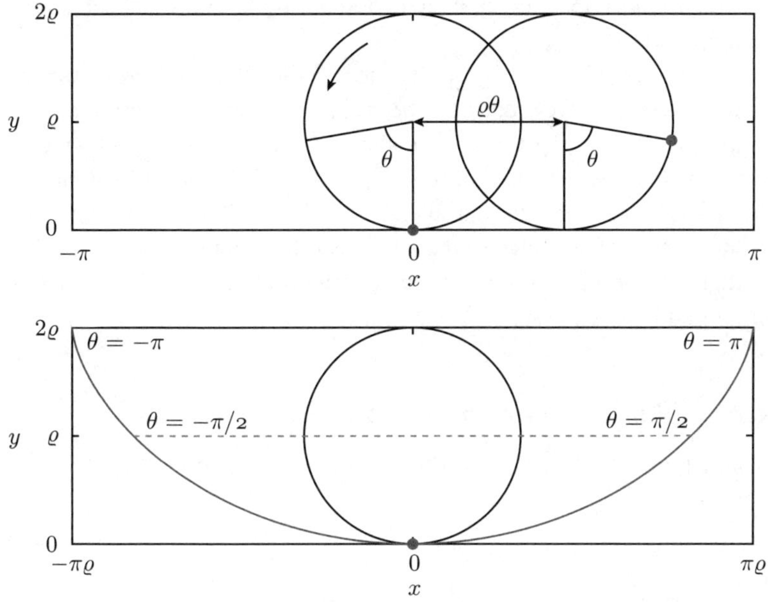

Figure 4.2. Construction of the cycloid where the wheel (generating circle of *radius* ϱ) rolls along the ceiling. Top: The rolling wheel starts with a downward valve and moves to the right: the cycloid then is traced by the valve. See the text for further explanation. Bottom: The cycloid as a curve in the (x, y)–plane, parametrized by the rolling angle θ.

4.2 The cycloid as isochrone and tautochrone curve

A few decades earlier Christiaan Huygens (1629–1695) had already been engaged with the cycloid [55], both as a curve and in connection to mechanics. In this section we now deal with these developments.

The right mechanical context is that of a bead that slides without friction along a given wire profile in a vertical plane, subjected to a constant vertical gravitational field. The bead then oscillates around a minimum in the wire profile and the question is for what profile these oscillations are *isochronous*, meaning that the period of oscillation is independent of its amplitude.

Again the solution is given by the cycloid. We shall prove this by showing that the cycloidal bead behaves as a *harmonic oscillator*. We cannot resist the tempation to also show that the cycloid is its own evolute. The historical context of this has to do with Huygens's solution of the isochronous pendulum problem, as well as the material realization of this with cycloidal cheeks mounted near the point of suspension. The background of this is formed by the problems around the determination of longitude at sea [87].

4.2.1 The harmonic oscillator

In general an oscillator is described by the equation of motion

$$\ddot{s} = -\frac{dV}{ds}(s), \tag{4.2}$$

where s is a one-dimensional position variable, think of a traversed distance (Latin: *spatium* = distance). With the expressions $\dot{s}$ and $\ddot{s}$ we indicate the first and second derivatives of s as a function of the time t, in other words the *velocity* and *acceleration* of the motion $s = s(t)$. Moreover V is the potential energy.

Remarks.

- The model is that the *restoring force* of the oscillator has the form $F = F(s)$, where Newton's (Isaac Newton 1642–1727) second law $F = m\,a$, with $a = \ddot{s}$ and where m is the point mass with position s, determines the motion of the oscillating particle.
- Introduction of $V(s)$ such that

  $$m\frac{dV}{ds}(s) = -F(s)$$

  then leads to the differential equation (4.2).

The function $V = V(s)$ is called the *potential energy* of (4.2).[1] For oscillation it is important that s varies near a minimum of V.

The oscillator is called *harmonic* when

$$V(s) = \frac{1}{2}\omega^2 s^2, \tag{4.3}$$

for a positive constant ω. The equation of motion (4.2) then becomes

$$\ddot{s} = -\omega^2 s, \tag{4.4}$$

which is a linear equation. The general solution [34, 53] as a function of the time t is given by

$$s(t) = R\cos(\omega t + \phi), \tag{4.5}$$

where the constants of integration R and ϕ are related to the initial conditions $s(0)$ and $\dot{s}(0)$ as follows:

$$s(0) = R\cos\phi \quad \text{and} \quad \dot{s}(0) = -\omega R\sin\phi.$$

That the solution (4.5) indeed satisfies (4.4) is easily checked. An important observation is that these solutions are all periodic with the same period

$$P = \frac{2\pi}{\omega}, \tag{4.6}$$

which therefore is independent of the amplitude R; we say that the harmonic oscillator is *isochronous*. The constant ω, the number of oscillations per unit of time, is called the *frequency* of the oscillator.

Harmonic oscillations in models. In physics the equation of motion (4.2) frequently occurs as a mathemematical model, often as an approximation of a nonlinear model. We here briefly mention a few examples, for more details compare [7, 25, 26, 38].

– **The vibrating spring.** We start by considering *vibrations of a spring*, which are the archetype for harmonic oscillations. See Figure 4.3 on

[1] For the moment leaving alone the precise physical dimensions.

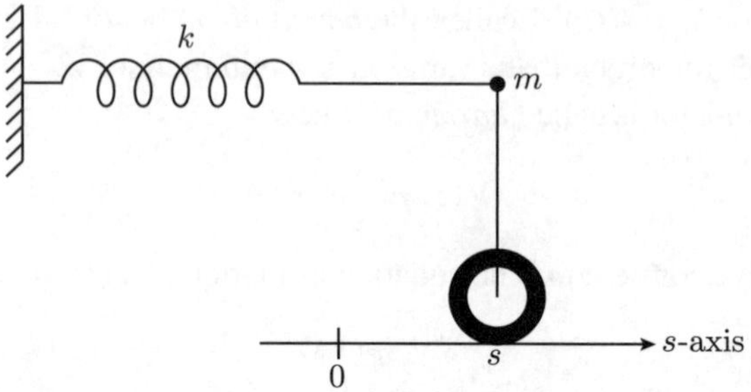

Figure 4.3. The harmonic oscillator as a spring, positioned horizontally in order to rule out gravitational influence. The position is given by the point where the wheel touches the s–axis.

page 74. To eliminate the influence of gravitation the spring is mounted horizontally. The spring itself has zero mass and at its tip we have a point mass m. Let s denote the one-dimensional position of this tip, where the value $s = 0$ corresponds to the situation without strain. By $s = s(t)$ we understand the position as a function of the time t, which describes the motion of the spring. The restoring force here is given by Hooke's law (Robert Hooke 1635-1703) i.e., by

$$F(s) = -ks,$$

in which k is the *spring constant*: k is large for stiff springs and small for soft ones. According to Newton's second law we now have

$$F(s(t)) = m\ddot{s}(t),$$

for all t. This yields the equation of motion

$$m\ddot{s} = -ks \tag{4.7}$$

for the spring. Writing

$$\omega = \sqrt{\frac{k}{m}},$$

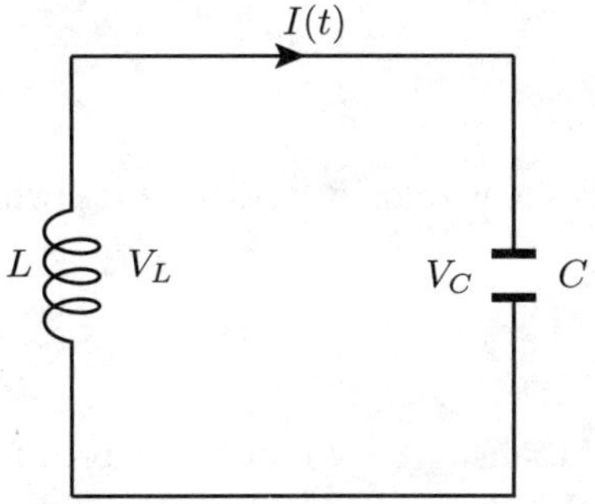

Figure 4.4. Electrical L-C network.

we see how (4.7) corresponds to (4.4), leading to the general solution (4.5).[2]

– **The electrical L-C network.** Now consider an electrical circuit containing a capacitor C and an inductor L, mounted as in Figure 4.4 on page 75. Assume that in the circuit we have a current I. Let the voltage differences over both parts be V_C and V_L respectively, then by Kirchhoff's Voltage Law (Gustav Robert Kirchhoff 1824-1887)

$$\mathrm{V}_C + \mathrm{V}_L = 0. \tag{4.8}$$

Moreover we know from the definitions of capacity and inductance [38] that

$$C\frac{d\mathrm{V}_C}{dt} = I \text{ en } \mathrm{V}_L = L\frac{dI}{dt}. \tag{4.9}$$

Differentiating (4.8) gives

$$\frac{d\mathrm{V}_C}{dt} + \frac{d\mathrm{V}_L}{dt} = 0.$$

Substitution of (4.9) for C and L further gives

$$\frac{I}{C} + L\frac{d^2I}{dt^2} = 0$$

[2]Note that in the formulas the physical dimensions may differ somewhat.

or

$$\ddot{I} = -\frac{1}{LC} I,$$

which is precisely the harmonic oscillator (4.4) with frequency

$$\omega = \sqrt{\frac{1}{LC}}.$$

Anharmonic oscillations. A notorious example of an anharmonic oscillator is the planar mathematical pendulum with equation of motion

$$\ddot{s} = -\omega^2 \sin s, \quad \omega = \sqrt{\frac{g}{\ell}}, \tag{4.10}$$

where ℓ is the length of the pendulum and g the gravitational acceleration. The position variable s is the angle of deviation, measured in radians. In Figure 4.5 a phase portrait is shown in the $(s, \dot{s})$-plane. The rest point $(0, 0)$ corresponds to the stable downward equilibrium, while the rest points at $(\pm\pi, 0)$ correspond to the unstable upside down equilibrium. The downward equilibrium is elliptic and surrounded by

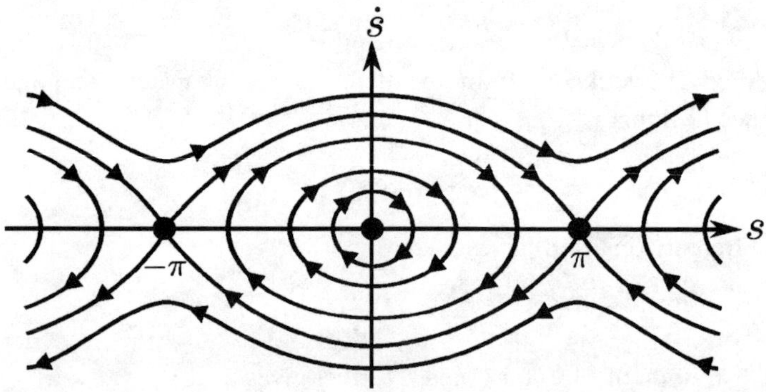

Figure 4.5. Phase portrait of the planar mathematical pendulum, depicted in the $(s, \dot{s})$-plane with the equilibria and a representative number of integral curves. The closed curves correspond to oscillatory motions of the pendulum. The curves at the top and bottom of the picture correspond to motions where the pendulum swings over the head.

oscillations. The upside down equilibrium is a saddle point. Since the motion is at rest in this equilibrium, nearby oscillations take ever longer. In fact, their period P tends to ∞ as the 'amplitude' tends to π.

Remarks.

- For small values of $|s|$ and $|\dot{s}|$ we may use the approximation $\sin s \approx s$, which again leads to the harmonic form (4.4). We speak of 'small oscillations' of the pendulum which seem to be isochronous. Folklore has it that Galileo (Galileo Galilei 1564–1642) had already established this as an experimental fact.
- Very well-known is the approximative formula [17]

$$P = 2\pi\sqrt{\frac{\ell}{g}}$$

  for the period of swing for small deviations and velocities, compare (4.6). This formula goes back to Huygens [55], also compare [3].

The sliding bead. From springs and pendulums we make a *Gestalt Switch* and think in terms of a wire profile in a vertical plane, along which a bead slides without friction under the influence of gravity. The pendulum, from this point of view, just amounts to a bead constrained to move on a *circular* wire; as mentioned above, this motion is not quite isochronous, only approximately so when the oscillations are small.

The problem now is for what wire profile does the bead oscillate isochronously. As said before, we shall see that the cycloid solves this problem. Our proof of this fact will rest on our considerations on the harmonic oscillator.

4.2.2 Intermezzo: the arclength of a cycloid

As we saw in Chapter 4, the cycloid with radius ϱ and rolling angle θ has parametrization

$$x(\theta) = \varrho(\theta + \sin\theta), \ y(\theta) = \varrho(1 - \cos\theta), \tag{4.11}$$

see Figure 4.6, also compare Figure 4.2 on page 71.

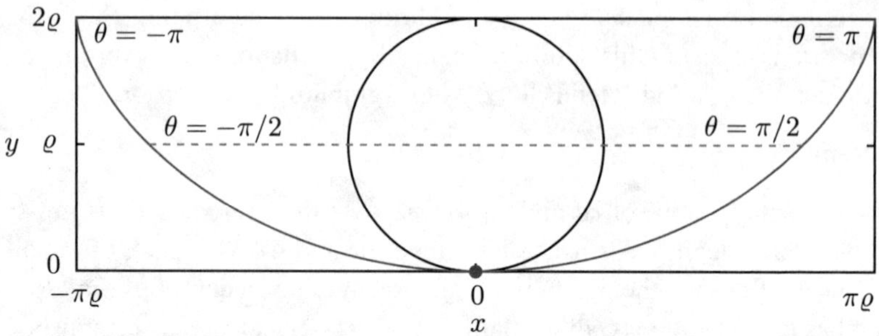

Figure 4.6. Cycloid with generating circle, with rolling angle θ and radius ϱ in the format of Figure 4.2.

The cycloid turns out to be *rectifiable*, in the sense that its arclength can be expressed in terms of elementary functions. Indeed, if $s = s(\theta)$ is the arclength, measured from the lowest point $\theta = 0$, we know by the Pythagorean Theorem

$$\begin{aligned} ds &= \sqrt{\left(\frac{dx}{d\theta}\right)^2 + \left(\frac{dy}{d\theta}\right)^2}\, d\theta = \\ &= \varrho\sqrt{2}\sqrt{1+\cos\theta}\, d\theta = 2\varrho\cos\left(\frac{1}{2}\theta\right)\, d\theta. \end{aligned}$$

Here note that for $-\pi < \theta < \pi$ we have that $\cos\left(\frac{1}{2}\theta\right) > 0$. Quadrature next leads to

$$s(\theta) = 4\varrho\sin\left(\frac{1}{2}\theta\right), \tag{4.12}$$

to be interpreted as arclength 'with sign'.

4.2.3 Dynamics along the cycloid

Here we show that the bead that slides along a cycloidal wire profile, as described above, indeed performs harmonic oscillations. The position of the bead on the cycloid from now on will be indicated by the arclength parameter $s = s(\theta)$, as in (4.12).

The main idea that the potential energy is proportional to the vertical height of the bead. When the bead is located at the point $(x(\theta), y(\theta))$ of the cycloid (4.1), then the vertical height equals $y(\theta)$, which depends of the arclength $s(\theta)$ as follows:

$$y(\theta) = 2\varrho \sin^2\left(\frac{\theta}{2}\right) = \frac{1}{8\varrho}(s(\theta))^2.$$

The potential energy now is given by

$$V(s) = \frac{g}{8\varrho}s^2,$$

where g is the gravitational acceleration. This means that the equation of motion (4.2) of the bead in terms of the position variable s is given by

$$\ddot{s} = -\frac{g}{4\varrho}s.$$

We recognize this as the harmonic oscillator (4.4) with frequency

$$\omega = \sqrt{\frac{g}{4\varrho}}, \tag{4.13}$$

with $\ell = 4\varrho$. This proves that the cycloid is isochronous, since for the period of oscillation we have

$$P = 2\pi\sqrt{\frac{4\varrho}{g}} = 4\pi\sqrt{\frac{\varrho}{g}},$$

independent of the amplitude.

Remarks.

- A representation of Huygens's proof [55] concerning isochrony can be found in Aarts [3].

- The cycloid also is the *tautochrone* curve: From any height y with $0 \le y \le 2\varrho$, the time to fall to the level $y = 0$ is equal to

$$T = \pi\sqrt{\frac{\varrho}{g}},$$

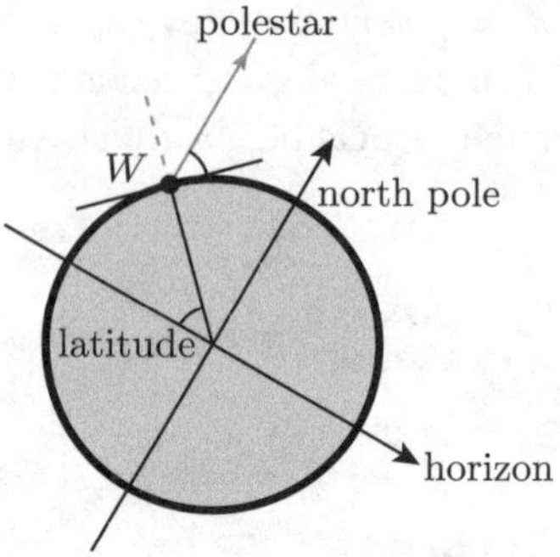

Figure 4.7. The north latitude equals the elevation of the polestar above the horizon.

for such a falling motion is nothing but one fourth of an oscillation, compare (4.2.3).

4.2.4 A historical digression

In the seventeenth and eighteenth centuries a major problem in navigation was the determination of the geographical longitude at sea. In a time without a Global Positioning System this could only be solved by studying the starry sky, armed with a sextant and with tables. The north latitude is not a problem since the altitude of the polestar above the horizon matches well with this, see Figure 4.7.[3]

The geographical longitude can be computed from the difference between the local time and, say, the Greenwich time. Twelve noon local time can be determined by shooting the sun at its maximum altitude while the Greenwich time must be kept by a precise clock. Since the earth rotates 360° in 24 hours, the difference in longitude in degrees of arc can be obtained from the time difference through multiplication by a factor of $15 = 360/24$.

However, in the times of Christiaan Huygens (1625–1695) only pendulum clocks were available, but they were not sufficiently reliable when

[3]On the southern hemisphere a similar construction exists based on the Southern Cross.

FIG.II.

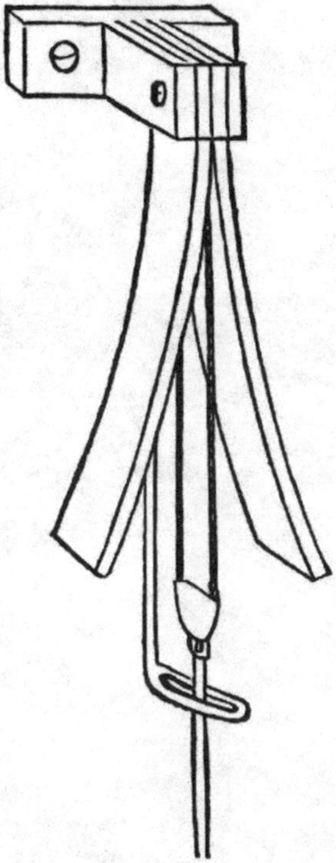

Figure 4.8. Cheeks mounted near the point of suspension of a pendulum clock. This picture is taken from Huygens's *Horologium Oscillatorium* [55].

used on ships. One problem is the an-isochronicity.[4] This means that the ordinary pendulum clock is unfit for maintaining Greenwich time over a longer time interval, say of weeks or months, with sufficient precision.

In this light Huygens asked himself whether there does exist an adapted pendulum which is isochronous. His idea was that the pendulum of the clock winds off along 'cheeks' that effectively shorten the

[4] Let alone the effects of the motion of the ship on the waves and in the wind.

Figure 4.9. Christiaan Huygens (1625–1695) by Caspar Netscher (ca. 1671), Haags Historisch Museum.

pendulum length and therefore also the period of swing at greater amplitude, see Figure 4.8 on page 81. The question then becomes what should be the form of these cheeks.

The solution to this problem forms the main subject of Huygens's monumental *Horologium Oscillatorium* [55]. This solution makes extensive use of properties of the cycloid. One of these is its isochrony and the specified question then becomes whether the cheeks can be designed in such a way that the pendulum mass will indeed trace a cycloid instead of a circle.

The answer is that the cheeks then also should be of cycloidal form: in fact they are segments of a cycloid that is congruent to the one that has to be traced. See Figures 4.10 (page 84) and 4.11 (page 86).

Determination of the geographical longitude at sea was a societal problem of great interest in a time without GPS. Armed with sextants and tables the sailors studied the starry sky and shot the sun. The latter occurred at 12 noon, local time – when the sun is at its highest altitude. Also coverings of stars by the moon were being observed. Much has been said about this, among others in the novel *Max Havelaar* [32] from the Dutch author Multatuli, by a fellow named Batavus Droogstoppel. On account of the essay *On the longitude at sea* in Sjaalman's package he says: "I think that at sea everything probably will be equally long as ashore." For more serious information see [87] and [95], also for a extensive historical survey.

Remarks.

- In museums such as the Royal Museum (Greenwich), Hofwijck (Voorburg), Teylers (Haarlem) and Boerhaave (Leiden), clocks can be seen with cycloidal cheeks near the point of suspension. Huygens also gave this kind of clockwork to captains to try out during their sea voyages.
- With the emergence of the spiral spring, that vibrates far more harmonically and therefore isochronously, the necessity to isochronize pendulum clocks disappeared. Such springs also suffer far less from the oscillatory motions of the ship. However, Huygens's mathematics and its considerable legacy remain highly interesting.

Below these thoughts will be explored further.

4.2.5 Two cycloids

We now explain the mathematics behind these cycloidal 'cheeks' along which the pendulum cord should wind off. Let C be the original cycloid from § 4.2.3 and consider also the congruent cycloid E mounted on top

ſit ferrea cuſpis DI, pauxillum ultra baſin inferiorem prominens, atque ita ut circumfe- DESCRIPTIO
rentiæ ejus exacte reſpondeat. HOROLOGII.

His ita ſe habentibus, ſi cylindrus ſecundum regulam AB volvatur, bracteolæ tan- (p. 11).
tum FG craſſitudine intercedente, eâque ſemper quantum poteſt extenſâ, deſcribet cuſpis I in ſubjecto tabulæ plano lineam curvam KI, quæ Cyclois vocatur. Circulus vero genitor erit CDE, cylindri adhibiti baſis. Quod ſi jam laminam KL ad regulam AB applicuerimus; exaratâ primum in ea cycloidis portione KI, invertemus deinde

[Fig. 19.]

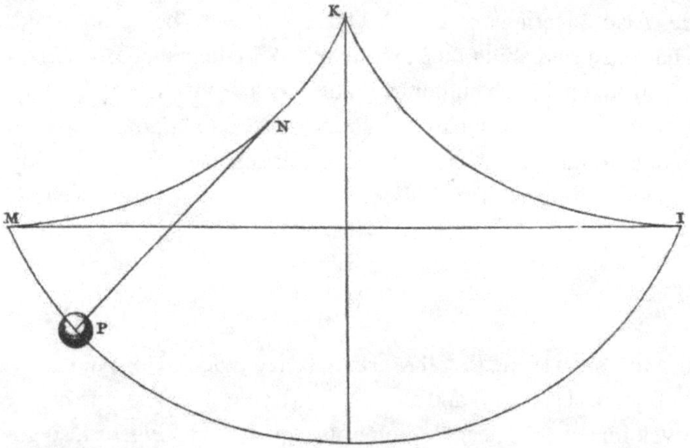

ipſam, & in ſuperficie adverſa ſimilem lineam KM, ab eodem puncto K egredientem, incidemus. Tum figuram MKI, accurate ſecundum lineas iſtas, efformabimus, cui figuræ lamellarum interſtitium aptari oportet, inter quas perpendiculum ſuſpenditur [1]). Sufficiunt autem ad horologiorum uſum portiones exiguæ arcuum KM, KI; reliquo flexu inutili futuro, ad quem perpendiculi filum accedere non poteſt.

Verum, ut mirabilis lineæ natura atque effectus plenius intelligantur, integras ſemicycloides KM, KI, alio ſchemate [Fig. 19] hic exprimere viſum fuit, inter quas ſuſpenſum agitatumque Pendulum KNP, diametri circuli genitoris duplum, cujuscunque amplitudinis oſcillationes, usque ad maximam omnium per arcum MPI, iiſdem temporibus confecturum ſit: atque ita, ut appenſæ ſphæræ P centrum, in linea MPI, quæ & ipſa cyclois integra eſt, ſemper verſetur. Quæ proprietas inſignis, neſcio an alii præter hanc lineæ data ſit, ut nempe ſe ipſam ſui evolutione deſcribat [2]). Hæc autem

14

Figure 4.10. The two cycloids according to Huygens [55].

of C as shown in Figure 4.11. Assume that a pendulum of length 4ϱ is fixed to the upper cusp point of E and allowed to swing in such a way that it touches the curve E, coming loose along the tangent line. Then the pendulum mass traces the lower cycloid C and it exactly describes the isochronous motion as described before in § 4.2.3.

To be precise let $C = C(\theta)$ be the cycloid (4.1), given by

$$x_C(\theta) = \theta + \sin\theta, y_C(\theta) = 1 - \cos\theta$$

and where for simplicity units are chosen in such a way that $\varrho = 1$.

The curve $E = E(\theta)$ then is parametrized as

$$\begin{aligned} x_E(\theta) &= \pi + (\theta - \pi) + \sin(\theta - \pi) \\ y_E(\theta) &= 2 + y_C(\theta - \pi), \end{aligned} \tag{4.14}$$

which simply rewrites to

$$x_E(\theta) = \theta - \sin\theta, \;\; y_E(\theta) = 3 + \cos\theta.$$

In both cases we set the parameter domain as $|\theta| \leq \pi$. Inspection of these formulas rapidly yields that E and C are two congruent cycloids, see Figure 4.11.

To investigate the above assertions on the relation between C and E, we consider the chord $K(\theta)$ between $C(\theta)$ and $E(\theta)$ and let its length be denoted by $k(\theta) = |K(\theta)|$, again see Figure 4.11. As an end point of E take the nearest tip at $\theta = \pm\pi$ and let $b(\theta)$ be the arclength of E, measured from $E(\theta)$ to the end point.

Proposition 4.1 (Confirmation of Huygens's idea). *For $|\theta| \leq \pi$ one has that $K(\theta)$ is tangent to E and perpendicular to C. Moreover $k(\theta) = b(\theta)$.*

Proof. The proof is just a matter of a careful check. On the one hand

$$\begin{aligned} K(\theta) = C(\theta) - E(\theta) &= \begin{pmatrix} 2\sin\theta \\ -2 - 2\cos\theta \end{pmatrix} \\ &= \begin{pmatrix} 4\sin\left(\frac{1}{2}\theta\right)\cos\left(\frac{1}{2}\theta\right) \\ -4\cos^2\left(\frac{1}{2}\theta\right) \end{pmatrix} = 4\cos\left(\tfrac{1}{2}\theta\right)\begin{pmatrix} \sin\left(\frac{1}{2}\theta\right) \\ -\cos\left(\frac{1}{2}\theta\right) \end{pmatrix}, \end{aligned}$$

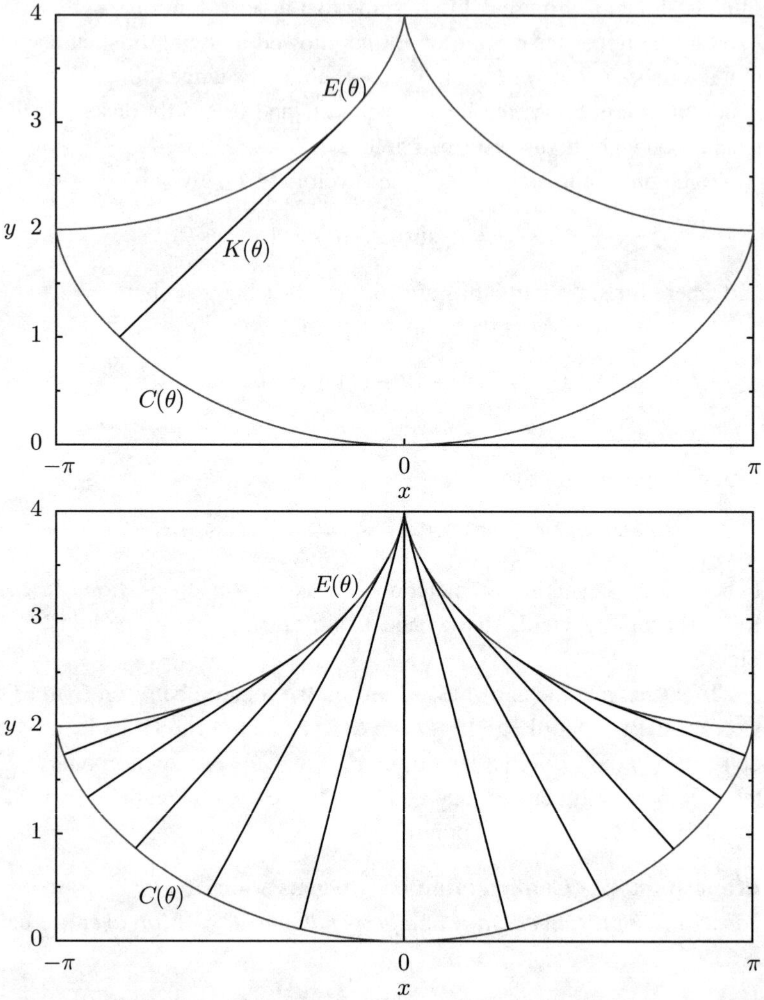

Figure 4.11. The cycloids E and C relate to each other as evolute and involute. See the text for explanation.

and on the other hand we have for any tangent vector in $E(\theta)$ to E

$$\begin{aligned} E'(\theta) &= \begin{pmatrix} 1-\cos\theta \\ -\sin\theta \end{pmatrix} \\ &= \begin{pmatrix} 2\sin^2\left(\frac{1}{2}\theta\right) \\ -2\sin\left(\frac{1}{2}\theta\right)\,\cos\left(\frac{1}{2}\theta\right) \end{pmatrix} = 2\sin\left(\tfrac{1}{2}\theta\right)\begin{pmatrix} \sin\left(\frac{1}{2}\theta\right) \\ -\cos\left(\frac{1}{2}\theta\right) \end{pmatrix}, \end{aligned}$$

from which it follows that $K(\theta)$ is parallel to $E'(\theta)$: the chord is tangent to the curve. The check that $K(\theta)$ is perpendicular to $C'(\theta)$ is completely analogous.

Finally we verify that $k(\theta) = b(\theta)$. On the one hand we have

$$k(\theta) = 2\sqrt{2}\sqrt{1+\cos\theta} = 4\cos\left(\tfrac{1}{2}\theta\right)$$

and on the other hand the arclength formula (4.12) gives

$$b(\theta) = 4\sin\frac{1}{2}(\theta+\pi) = 4\cos\left(\tfrac{1}{2}\theta\right). \qquad \square$$

Remarks.

- That $K(\theta)$ is perpendicular to C has everything to do with the fact that development of the cord at the point $E(\theta)$ amounts to an 'instantaneous' rotation.
- Perpendicularity further means that the swinging motion exactly coincides with that of the bead from § 4.2.3, because the force exerted by the cord after all does not perform any work [7].[5]

In summary, the point of suspension of the pendulum cord is the 'cusp' in E, and the cord develops along the cycloidal cheeks E, coming off along the tangent K as shown in Figure 4.11. The pendulum mass then follows the curve C in an isochronous motion.

[5] A similar remark applies for the circular motion of the ordinary pendulum.

4.2.6 Scholium

In later developments of geometry Huygens's ideas were picked up and further elaborated in the notions of *evolute* and *involute*, see, e.g., Do Carmo [30].

Definitions. The general construction assumes a parametrized curve $C = C(\theta)$, but it helps to think of the cycloid (4.1) and meanwhile to inspect Figure 4.11. The *evolute* E of C now is the set of centers of curvature of C. Here, the *center of curvature* $E(\theta)$ of a point $C(\theta)$ is the center of the circle that at this point has maximal contact with C.[6] This is the so-called *osculating* circle. The radius of the osculating circle is the *radius of curvature* in $C(\theta)$ and the *curvature* in $C(\theta)$ is exactly the (multiplicative) inverse of the radius. From this construction it follows that E is the *envelope* of the normal bundle of C and we say that C is an *involute* of E.[7]

Vice versa we say that E is an *evolute* of C, defined in the following way. Take an end point of the curve E. At any point P on E we consider the tangent to E. The points Q on these tangents, such that the distance $|P - Q|$ equals the arclength between P and the given endpoint, then form an involute C. Generally, E is an evolute of C if C is an involute of E with respect to some endpoint. Compare [2, 3].

Remarks.

- Lagrange [60] (Joseph Louis Lagrange 1736–1813) found the isochrone curve in a more systematic way using a differential equation. He too was looking for the harmonic oscillator.

- From Proposition 4.1 we see that E is tangent to all normals $K(\theta)$ on the curve C, also see Figure 4.11, top and bottom. In other words this means that E envelopes the normal bundle of the curve C, which has another optical interpretation. Here C is a source of light and E a *caustic* formed by the ensuing light rays. Also see Chapter 8.

[6]This means that in $C(\theta)$ the 0th, 1st and 2nd derivatives of the circle coincide with those of C.

[7]We also speak of the involute C as a *developing curve*, sometimes also called *evolvent*.

A question that may well occur to the reader is how Huygens [55] found all of this and even more without having the calculus at his disposal. For a clarification of all this see Aarts et al. [3] and also Whitman [97] and Yoder [98]. For more general background information we are happy to refer to Andriesse [6].

4.3 The flat earth atmosphere revisited

Returning to the brachistochrone problem, we have considered the motion of the bead K under the influence of constant gravity, again see Figure 4.1 on page 12. When the bead is at rest at the highest point A, its velocity increases until it reaches the lowest point B. By conservation of energy its potential energy decreases while its kinetic energy increases, in such a way that their sum

$$M = \frac{1}{2}m\ (v(y))^2 + mgy, \tag{4.15}$$

is constant during the entire falling motion. Here m is the mass of the bead K, g is the acceleration of gravity, v is the velocity and M the total energy. Thus the (total) energy is conserved during the motion. The fact that the velocity v of the bead depends only on the vertical height y suggests an analogy to the atmosphere of a flat earth as in Chapter 2. So we can model the situation of the bead by thinking of it as a ray of light whose velocity at height y is the one given by the gravational field. This leads to the choice

$$n(y) = \frac{1}{v(y)}$$

for the refraction index profile.

4.3.1 Two conservation laws

Besides the conservation of energy (4.15), in the optical setting we have another conserved quantity, namely (2.8)

$$S = n(y)\sin\alpha,$$

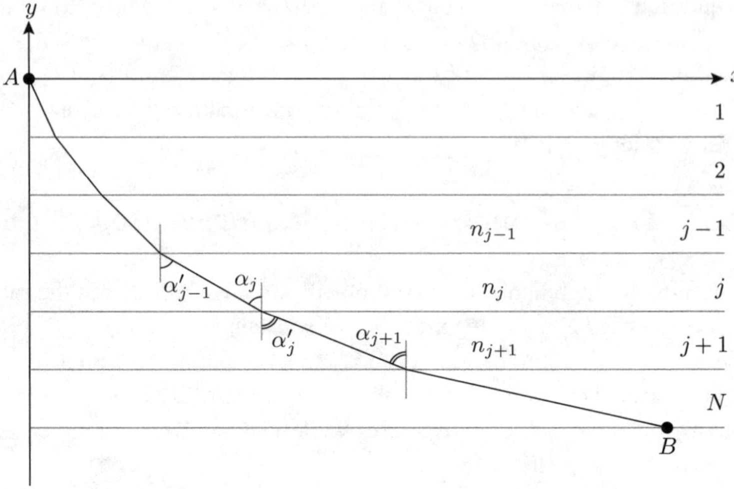

Figure 4.12. Bernoulli's optical set-up. Similar discrete approximations of a smooth situation are characteristic for applications of differential and integral calculus as also already visible in the work of Newton and Leibniz.

where α again is the inclination with the vertical direction, compare Figure 4.12 on page 90. The conserved quantity S is related to the name 'Snell'. Also compare Corollary 2.2 and Figure 2.6 (page 28).

4.3.2 The cycloid as the brachistochrone curve

As we saw at the end of § 2.3.1, the conservation law (2.8) determines the form of the light ray, in this case of the brachistochrone. It turns out to be more convenient not to express this curve in the form $x = x(y)$, but rather to parametrize it as $x = x(\alpha)$, $y = y(\alpha)$. From the conservation laws (2.8) and (4.15) we now prove the following theorem.

Theorem 4.2 (Brachistochrone as cycloid). *Assuming the above circumstances, the brachistochrone curve has the cycloidal form*

$$\begin{aligned} x(\alpha) &= \frac{1}{4S^2 g}(2\alpha - \sin(2\alpha) \\ y(\alpha) &= \frac{1}{4S^2 g}(-1 + \cos(2\alpha)), \end{aligned} \tag{4.16}$$

passing through $A = (0, 0)$ for $\alpha = 0$ and where the constant S determines the scale of the curve such that it passes through B.

Remarks.

– In terms of the format (4.1) and Figure 4.2 (page 71), the equation (4.16) describes a cycloid with *rolling angle* θ and *radius* ϱ given by

$$\theta = 2\alpha - \pi \quad \text{and} \quad \varrho = \frac{1}{4S^2 g}.$$

– The cycloid played an important role in the 17th century, in fact it was a 'much talked-of curve', see [50, 71, 80]: it is even compared to Helen of Troy. After the fundamental work of Huygens, Bernoulli's brachistochrone result came as a surprise.

Proof. Since the motion is pointing down, we have $dx/dy = -\tan\alpha$. We derive the theorem from the conservation laws (2.8) and (4.15)

$$S = n(y)\sin\alpha \quad \text{and} \quad M = \frac{1}{2} m\, v(y)^2 + mgy\,.$$

As mentioned before the desired profile will be of the form

$$(x(\alpha), y(\alpha)), \tag{4.17}$$

where α is the inclination with the y–direction.

Since M is constant, differentiating the second equation we get that for all y we have

$$v'(y) = -\frac{g}{v(y)}\,. \tag{4.18}$$

Rewriting the conservation (2.8) as $\sin\alpha = Sv(y)$, differentiation with respect to α leads to

$$\cos\alpha = Sv'(y)\,\frac{dy}{d\alpha}\,. \tag{4.19}$$

Next we determine the derivatives $dy/d\alpha$ and $dx/d\alpha$ for the parametrized format (4.17). The results thus obtained are suitable for integration.

First we consider $dy/d\alpha$:

$$\begin{aligned}\frac{dy}{d\alpha} &\overset{(4.19)}{=} \frac{\cos\alpha}{Sv'(y)}\\ &\overset{(4.18)}{=} -\frac{v(y)}{Sg}\cos\alpha\\ &\overset{(2.8)}{=} -\frac{1}{2S^2g}\sin(2\alpha)\end{aligned} \tag{4.20}$$

and next $dx/d\alpha$:

$$\begin{aligned}\frac{dx}{d\alpha} &= -\tan\alpha\,\frac{dy}{d\alpha}\\ &\overset{(4.20)}{=} \frac{v(y)}{Sg}\sin\alpha\\ &\overset{(2.8)}{=} \frac{1}{2S^2g}(1-\cos(2\alpha)).\end{aligned}$$

Integration now gives

$$\begin{aligned}y(\alpha) &= C_y + \frac{1}{4S^2g}\cos(2\alpha)\\ x(\alpha) &= C_x + \frac{1}{4S^2g}(2\alpha - \sin(2\alpha),\end{aligned}$$

where we choose the constants of integration C_y and C_x in such a way that $x(0) = 0 = y(0)$, which leads to $C_y = -1/(4S^2g)$ and $C_x = 0$. So we finally get

$$\begin{aligned}x(\alpha) &= \frac{1}{4S^2g}(2\alpha - \sin(2\alpha)\\ y(\alpha) &= \frac{1}{4S^2g}(-1+\cos(2\alpha)),\end{aligned}$$

as claimed. □

Remarks.

– The value of the energy M is not directly visible in the solution (4.16), but in the constant of integration this information comes back:

$$M = \frac{1}{2}m\,(\upsilon(y_0))^2 + mgy_0.$$

Since we assume that $\upsilon(y_0) = 0$ the bead starts at a height $y = y_0$ at rest. It then follows that during the entire motion we have $M = mgy_0$ and the corresponding falling speed therefore equals $\upsilon(y) = \sqrt{2g(y_0 - y)}$, where $y \leq y_0$.
By a translation on the y–axis we may even take $M = 0$. In that case the falling speed is given by

$$\upsilon(y) = \sqrt{-2gy},$$

for $y \leq 0$. The kind of consideration is rather customary in classical mechanics.

– Direct integration of $x = x(y)$ according to Theorem 2.3, in particular of (2.10), yields formulas as, for instance, described in [4], § 3.3 *Integrals of Irrational Algebraic Functions*; also [78]. It is a what Mark Levi calls a 'finger-breaking' exercise [68] to reconcile the above results with the direct approach from § 4.3.2 based on parametrization by the inclination α.

– Notice that the brachistochrone cannot be expressed as a univalent function $x = x(y)$, which is a consequence of the 'reflection' that occurs at $y = -2\varrho$ in Figure 4.12. In Chapter 7 more generic forms of reflection will be studied.

4.4 Scholium

In the seventeenth century the cycloid played a special role in the development of differential and integral calculus. In the work of Johann Bernoulli the brachistochrone curve saw the light of day in 1696 as a contest in the journal *Acta Eruditorum* [14]. One year later he published his own vent s'étendre plus amplement vers en haut, & moins vers en bas, mais vers les autres endroits plus ou moins ſelon qu'ils

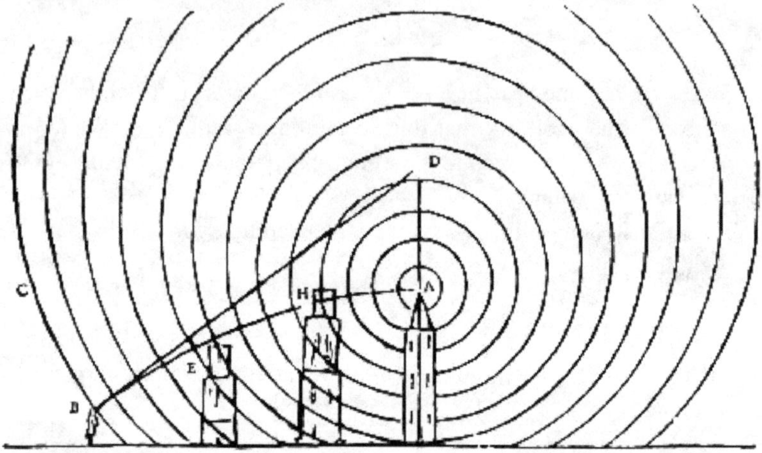

approchent de ces deux extremes. Ce qui eſtant, il s'enſuit neceſſairement que toute ligne, qui coupe une de ces ondes à angles droits, paſſe au deſſus du point A, ſi ce n'eſt la ſeule qui eſt perpendiculaire à l'horizon.

Soit B C l'onde qui porte la lumiere au ſpectateur qui eſt en B, & que B D ſoit la droite qui coupe cette onde perpendiculairement. Or parce que le rayon ou la ligne droite, par laquelle nous jugeons l'endroit où l'objet nous paroit, n'eſt autre choſe que la perpendiculaire à l'onde qui arrive à noſtre oeil, comme l'on peut entendre par ce qui à eſté dit cy deſſus, il eſt manifeſte que le point A s'appercevra comme eſtant dans la droite B D, & ainſi plus haut qu'il n'eſt en effet.

De meſme ſi la Terre eſt A B, & l'extremité de l'Atmoſphere

C D

Figure 4.13. A light ray in a cycloidal form by Huygens [56]. For a discussion see the contribution on Johann Bernoulli in [86]. In his commentary on the brachistochrone problem the latter says that although Huygens did have the right figure, he had no explicit formula [14].

solution in the same journal. A number of his famous contemporaries quickly responded, Leibniz being one of them. Newton's solution was published under a pseudonym, but Bernoulli "recognized the lion by his paw."[8] For historical details see Goldstine [44]. The brachistochrone problem and its solutions mark the birth of the long and broad development of the Variational Principle, that nowadays forms a cornerstone of many scientific disciplines. This boils down to the idea that nature behaves in a way that corresponds to solving an optimization problem. For background on the ensuing variational calculus and the Bernoulli family, see [44, 70, 85, 91] and also [62, 63, 64], as well as all their bibliographies.

> Leibniz firmly believed in a large general validity of the Variational Principle: he says somewhere with so many words that our world is the best of all possible ones. Voltaire has immortally ridiculed this in his famous book *Candide, ou l'optimisme* [13]. Nevertheless this way of thinking, where the Bernoulli's had a leading role, had a tremendous influence on the development of science.

In Chapter 6 the brachistochrone will again appear on the stage; one of the themes of this book is the development of variational calculus in optics, where the brachistochrone is a nice leading example. Compare [85, 91].

[8] . . . ex ungue leonem

II

Light rays as geodesics

Chapter 5

The principles of Fermat and Hamilton

Part I of this book, particularly its mathematical heart Chapter 2, uses only some differential and integral calculus and some Euclidean geometry. In Part II we enter more deeply into the mathematical background of geometric optics, introducing a few elements from the 18th and 19th century theory of variational calculus and differential geometry. The prerequisites therefore also have been increased somewhat. See [30, 82, 88] for general references on differential geometry. This excursion provides us with tools for a better understanding of the optical phenomena in the atmosphere.

We start by introducing the differential geometrical setup, in which light rays follow geodesics; this turns out to be a suitable and elegant translation of the Fermat Principle. Here minimization of the time is translated into the minimization of a certain kind of distance, to be expressed in terms of a Riemannian metric. In such a metric the curves of shortest distance are known as geodesics. These geodesics then can be found using the variational calculus of Euler and Lagrange, which is based on earlier work of, among others, the Bernoullis and Leibniz. The resulting differential equations are known as the *eikonal equations* of geometric optics. The 'final' translation of the Fermat Principle in terms of variational calculus was performed by Hamilton. We shall therefore speak of Hamilton's Principle.

5.1 From Fermat via Euler-Lagrange to Hamilton

For a given smooth curve $\tau \mapsto q(\tau)$ in the plane or in space we denote the velocity vector $\dot{q}$ by $\dot{q} = dq/d\tau$. At this moment we do not specify the parameter τ, since it turns out that the theoretical background will provide us with a very natural choice for τ. If this curve parametrizes a light ray in an isotropic medium, the refraction index $n(q)$ is well-defined. By the Chain Rule we know

$$\frac{dq}{dt} = \frac{d\tau}{dt}\frac{dq}{d\tau}. \tag{5.1}$$

Here t denotes the time. This is vectorial notation, but if q is a plane curve and we write $q(\tau) = (x(\tau), y(\tau))$ then (5.1) just becomes the two scalar equations

$$\frac{dx}{dt} = \frac{dx}{d\tau}\frac{d\tau}{dt}$$
$$\frac{dy}{dt} = \frac{dy}{d\tau}\frac{d\tau}{dt},$$

which is the Chain Rule in a more familiar format. The velocity vector is just

$$\frac{dq}{d\tau}.$$

If we now use that $\|d\,q(t)/dt\| = 1/n(q(t))$, by definition of the refraction index, then a short calculation yields

$$dt = n(q)\|\dot{q}\|\,d\tau. \tag{5.2}$$

In these terms, the Fermat Principle says that for $\tau_1 < \tau_2$, where $A = q(\tau_1)$ and $B = q(\tau_2)$ are kept fixed, the integral

$$\int_{\tau_1}^{\tau_2} dt = \int_{\tau_1}^{\tau_2} n(q(\tau))\|\dot{q}(\tau)\|\,d\tau$$

should be minimal (or at least extremal) under small variations of the curve q. Since it is not convenient to deal with square roots, we rather

Figure 5.1. Sir William R. Hamilton (1805–1865) and G. F. Bernhard Riemann (1826–1866)

extremize the integral

$$\mathcal{I}(q) = \int_{\tau_1}^{\tau_2} \frac{1}{2} n^2(q(\tau)) \|\dot{q}(\tau)\|^2 \, d\tau \tag{5.3}$$

as a function of the curve q; locally this amounts to the same problem. The integrand of (5.3)

$$L(q, \dot{q}) = \frac{1}{2} n^2(q) \|\dot{q}\|^2 \tag{5.4}$$

(where $n^2(q) = (n(q))^2$) is known as the *Lagrangian* of this minimization problem; it can also be interpreted as the (kinetic) energy of the present motion. The reformulation of the Fermat Principle now says that light rays extremize the integral (5.3). This is precisely what is known as Hamilton's Principle [47].

Remarks.

- In the case of the Lagrangian (5.4) the extremum of the functional (5.3) is in fact a minimum, at least as long as $\tau_2 - \tau_1$ is sufficiently small.

Compare similar remarks following Theorems 2.1 and 2.4. For a general discussion on large scale problems see [42, 84].

- The introduction of the parameter τ may look somewhat mysterious at first, and one may well be tempted to simply choose $\tau = t$, but it turns out not to be wise to make an early choice. In fact, the underlying differential geometry is going to provide us with a natural 'geodesic' parametrization.

5.1.1 Light rays as geodesics

In the formula $dt = n(q)\|\dot{q}\|\, d\tau$, see (5.2), we interpret the *time* as a *distance*. In effect, we treat distances in the optical medium as dependent on the refraction index: locally, distances become longer when the refraction index n gets larger. This suggests the introduction of a Riemannian metric. This is an inner product on the tangent space that depends on q, as follows:

$$G_q(\dot{q}_1, \dot{q}_2) = n^2(q)\langle \dot{q}_1, \dot{q}_2 \rangle\,, \tag{5.5}$$

where $\langle \dot{q}_1, \dot{q}_2 \rangle$ denotes the standard (Euclidean) inner product of the tangent vectors $\dot{q}_1$ and $\dot{q}_2$. That G is a scalar multiple of the standard inner product reflects the isotropy of the medium: at any point q, the speed of light in all directions has the same value $1/n(q)$. From this it follows that

$$L(q, \dot{q}) = \frac{1}{2} G_q\langle \dot{q}, \dot{q} \rangle\,, \tag{5.6}$$

compare (5.4). Since geodesics locally minimize distance, in this setting extremization of the integral $\mathcal{I}$ in (5.3), by definition, exactly yields the geodesics of the Riemannian metric G. From this we may conclude that the light rays, as obtained by the Hamilton Principle, are indeed the geodesics of this Riemannian metric. This provides us with a more modern entrance to geometric optics [10, 46].

In Chapter 7 the geometric optics of the rotationally-symmetric atmosphere of the round earth will be extensively discussed in terms of the differential geometry of a surface of revolution in three-dimensional space.

Remarks.

- The extremum of $\mathcal{I}$ in our case locally is a minimum as long as $\tau_2 - \tau_1$ is sufficiently small. That this holds in general only locally is a well-known fact in differential geometry. A famous example thereof concerns the standard 2–sphere $\mathbb{S}^2$, on which the geodesics are *great circles*, whose centers coincide with the center of the sphere. If, as after Theorem 2.4, we consider two points A and B, where B moves away from A along a great circle, then the connecting arc is minimal as long as B has not yet reached the antipodal point of A. After this the other arc along the same great circle becomes the shortest connection. We say that A and its antipodal point are *conjugated*, which here expresses as much as that (all) geodesics starting at A meet again at B.

5.1.2 Euler and Lagrange

Leonhard Euler (1707–1783) was a student of Johann Bernoulli (1667–1748) and Joseph Louis Lagrange (1736–1813) was a student of Euler. Both made many fundamental contributions to mathematics and (mathematical) physics. Regarding the calculus of variations, Euler improved on the Bernoullis and on Leibniz, while Lagrange in his turn improved on Euler. Their joint efforts resulted in the Euler–Lagrange equations, which provide extremal solutions for the problem to extremize the functional

$$\mathcal{I}(q) = \int_{\tau_1}^{\tau_2} L(q(\tau), \dot{q}(\tau))\, d\tau\ .$$

as in (5.3), for very general Lagrangians L. We here briefly summarize this theory, referring to [7, 43] and the corresponding bibliographies for details. First we state the following theorem without proof.

Theorem 5.1 (Euler–Lagrange [36, 61]). *Assume that $n = n(q)$ is a smooth function. Then a curve $\tau \in \mathbb{R} \mapsto q(\tau) \in \mathbb{R}^2$ satisfies the* Euler–Lagrange equations

$$\begin{aligned} \frac{d}{d\tau}\frac{\partial L}{\partial \dot{q}_1} &= \frac{\partial L}{\partial q_1} \\ \frac{d}{d\tau}\frac{\partial L}{\partial \dot{q}_2} &= \frac{\partial L}{\partial q_2} \end{aligned} \tag{5.7}$$

if and only if the integral $\mathcal{I}$ of (5.3) *is extremal under small variations of the curve q, where the end points $q(\tau_1)$ and $q(\tau_2)$ are kept fixed.*

Remarks.

- To illustrate how this works we give a brief discussion of the Euler–Lagange equations (5.7) in the optical setting of the (x, y)–plane with refraction index profile $n = n(y)$. In the above we take $(q_1, q_2) = (x, y)$. For the computation of the partial derivatives of the Lagrangian $L(x, y, \dot{x}, \dot{y}) = \frac{1}{2}n^2(y)\left((\dot{x})^2 + (\dot{y})^2\right)$ we have to consider x, y, $\dot{x}$ and $\dot{y}$ as independent variables. This leads to

$$\frac{\partial L}{\partial x} = 0, \quad \frac{\partial L}{\partial y} = n(y)n'(y)\left((\dot{x})^2 + (\dot{y})^2\right)$$

  and

$$\frac{\partial L}{\partial \dot{x}} = n^2(y)\dot{x}, \quad \frac{\partial L}{\partial \dot{y}} = n^2(y)\dot{y},$$

  where $n' = dn/dy$.

  Substitution of these expressions into (5.7) gives a system of second order differential equations

$$\begin{aligned} &\frac{d}{d\tau}\left(n^2(y)\dot{x}\right) = 0 \\ &\frac{d}{d\tau}\left(n^2(y)\dot{y}\right) = n(y)n'(y)\left((\dot{x})^2 + (\dot{x})^2\right), \end{aligned} \tag{5.8}$$

  where now $\dot{x} = dx/d\tau$ and $\dot{y} = dy/d\tau$.

- Effectively performing the differentiations in (5.8) becomes difficult. In a while we shall see how Hamilton's approach is a considerable simplification, where both second order differential equations are replaced by two first order equations. However, one conclusion we can already draw at this moment, is that the expression $n^2(y)\dot{x}$ is a conserved quantity in (5.8). We shall elaborate on this later.

5.1.3 Via Legendre to Hamilton

By the Legendre transformation (Adrien-Marie Legendre 1752-1833) an elegant transfer can be made from Euler–Lagrange to the *canonical* formalism of Hamilton.

Theorem 5.2 (Hamilton [7, 43]). *Consider the* Legendre transformation $\mathcal{L} : \mathbb{R}^4 \longrightarrow \mathbb{R}^4$

$$\begin{aligned} \mathcal{L} : (q_1, q_2, \dot{q}_1, \dot{q}_2) &\mapsto (q_1, q_2, p_1, p_2), \quad \textit{where} \\ p_j = \frac{\partial L}{\partial \dot{q}_j} &= n^2(q)\,\dot{q}_j \end{aligned} \tag{5.9}$$

for $j = 1, 2$.

The transformation $\mathcal{L}$ *is invertible and the* Hamiltonian

$$H : \mathbb{R}^4 \longrightarrow \mathbb{R}, \ \textit{where}\ H(p, q) = L \circ \mathcal{L}^{-1}(q, p) = L\left(q, \frac{p}{n^2(q)}\right) \tag{5.10}$$

has the form

$$H(p, q) = \frac{1}{2n^2(q)} \|p\|^2. \tag{5.11}$$

The solutions of the Euler–Lagrange equations (5.7) *correspond with those of the* canonical *equations*

$$\begin{aligned} \dot{q}_j &= \frac{\partial H}{\partial p_j} \\ \dot{p}_j &= -\frac{\partial H}{\partial q_j}, \end{aligned} \tag{5.12}$$

for $j = 1, 2$.

Proof. From (5.9) it directly follows that for the inverse transformation $\mathcal{L}^{-1}$ we have

$$\dot{q}_j = \frac{1}{n^2(q)} p_j,$$

$j = 1, 2$. The general theory [7, 43] that first of all provides us with the form (5.10), which, together with (5.4) for the Lagrangian L, directly leads to (5.11). Moreover we note that the Euler–Lagrange equations (5.7) by definition (5.9) become equivalent to the system

$$\dot{p}_j = \frac{\partial L}{\partial q_j}$$

for $j = 1, 2$. □

Remarks.

- The canonical equations (5.12) are often called the *Hamilton–Jacobi* equations (Carl Gustav Jacobi 1804-1851).
- In a more geometric setup, $\mathcal{L}$ is a transformation from the tangent bundle to the co-tangent bundle of the configuration space with coordinates q_j that correspond to the positions of the light rays. In the background of Theorem 5.2 it is relevant that the Lagrangian L in the tangent space attached to q is a positive definite quadratic form in the $\dot{q}_j$–variables.

5.1.4 General remarks

At the conclusion of this section we present a few remarks concerning the systems (5.7) and (5.12) and their applications in geometric optics. In the sequel we shall pass back and forth at will between these equivalent formulations.

As said earlier L, and therefore also H, stands for the total energy of the systems (5.7) and (5.12).

Theorem 5.3 (Conservation of Energy). *Assuming the circumstances of Theorem 5.2, along any integral curve $\tau \mapsto (q(\tau), \dot{q}(\tau))$, and, equivalently, along $\tau \mapsto (q(\tau), p(\tau))$, the energy*

$$L(q(\tau), \dot{q}(\tau)) \equiv H(q(\tau), p(\tau)) \tag{5.13}$$

is constant.

Proof. We use vectorial notation. For the directional derivative $\dot{H}$ of the function H along the vector field given by (5.12) we have

$$\begin{aligned}\dot{H} &= \frac{\partial H}{\partial p}\dot{p} + \frac{\partial H}{\partial q}\dot{q} \\ &= -\frac{\partial H}{\partial p}\frac{\partial H}{\partial q} + \frac{\partial H}{\partial q}\frac{\partial H}{\partial p} \\ &\equiv 0,\end{aligned}$$

which establishes our claim. □

For a given value E conservation of energy can be expressed as

$$L(q(\tau), \dot{q}(\tau)) \equiv E \equiv H(q(\tau), p(\tau)). \tag{5.14}$$

Note that in terms of (5.5) this exactly means that

$$\langle \dot{q}, \dot{q} \rangle_G = 2E. \tag{5.15}$$

In the case where $2E = 1$ we see that the geodesic is parametrized according to arclength.

Remarks.

- For a given refraction index profile $n = n(q)$ the differential equations (5.7),[1] determine the ray bundles in a unique way. Compare Theorem 5.2.

- Without further effort q here can stand for a three-dimensional position variable $q = (q_1, q_2, q_3)$. The essential formulas (5.4) and (5.11), as well as the corresponding differential equations, would not change, certainly not in the present vectorial form. In this way we get a general entrance to geometrical optics, without the restriction to a vertical plane [9, 38, 46].

- Evidently, the differential equations (5.7) and (5.12) also give the possibility of approximating the corresponding ray bundles numerically

[1] Also known as the *eikonal-equations* of geometric optics.

for any given refraction index profile. As we said before in Chapter 3, the inverse problem is of particular interest: to determine an appropriate refraction index profile for given optical observatons, compare [65, 95].

5.2 The Noether Principle: Conservation laws by symmetry

A fairly general way to obtain conserved quantities runs as follows. One of the variables in $q = (q_1, q_2)$, say q_1, does not explicitly show up in the expression for the Lagrangian L, see (5.4). We say that q_1 is a *cyclic variable*. This is a special way for the Noether Principle to work: q_1 being cyclic expresses that the system is symmetric under translations in the q_1-direction. From this it follows that

$$\frac{\partial L}{\partial q_1} \equiv 0, \tag{5.16}$$

which in (5.7) leads to

$$\frac{d}{d\tau}\frac{\partial L}{\partial \dot{q}_1} \equiv 0\,,$$

from which we see that

$$p_1 = \frac{\partial L}{\partial \dot{q}_1} \tag{5.17}$$

is a conserved quantity. Also in the Hamiltonian formulation this relation can be clearly seen. Indeed, in (5.12) one has

$$\dot{p}_1 = -\frac{\partial H}{\partial q_1} \equiv 0, \tag{5.18}$$

from which we immediately get the same result, namely that p_1 is conserved. We shall apply this idea both for the flat earth atmosphere with $q_1 = x$, $q_2 = y$ and for the round earth eatmosphere with $q_1 = \varphi$, $q_2 = r$.

Amalie Emmy Noether (1882–1935) was a mathematician, who among other things, became famous for the discovery that in the context of the differential equations (5.7) and (5.12) a wonderful relation exists between symmetries and conservation laws. In it simplest formulation we first observe that a conserved quantity F generates a vector field

$$X_F = \frac{\partial F}{\partial p}\frac{\partial}{\partial q} - \frac{\partial F}{\partial q}\frac{\partial}{\partial p}.$$

This notation abbreviates (5.12), with H replaced by F. The solution flow generated by X_F forms a symmetry group of the differential equations (5.7) and (5.12). The other way around, given a symmetry group one can search for an infinitesimal generator, which then is a conserved quantity. The Noether Principle is a theorem that guarantees that all this works well under quite general circumstances. For a textbook version, see [7].

5.2.1 The flat earth atmosphere

We now inspect what the Noether Principle means for the geometric optics of the flat earth atmosphere. In this case $q = (x, y)$ and the horizontal translation symmetry is expressed by the fact that the refraction index profile has the form $n = n(y)$, i.e., independent of x. According to (5.4) we now get

$$L(q, \dot{q}) = \frac{1}{2}\langle \dot{q}, \dot{q}\rangle_G = \frac{1}{2}n^2(y)\left(\dot{x}^2 + \dot{y}^2\right) \tag{5.19}$$

and clearly x is indeed a cyclic variable and therefore

$$p_x = n^2(y)\,\dot{x} \tag{5.20}$$

is a conserved quantity. Also compare the the example in § 5.1.2. Bearing in mind the mechanical analogy [7, 43], the function p_x is known as the *momentum*.

Also introducing $p_y = n^2(y)\,\dot{y}$ as in (5.9), application of the above to (5.11) for this case yields

$$H(p, q) = \frac{1}{2n^2(y)}(p_x^2 + p_y^2)\,. \tag{5.21}$$

The corresponding canonical equations (5.12) therefore read

$$\begin{aligned} \dot{x} &= \frac{\partial H}{\partial p_x}, \quad \dot{p}_x = -\frac{\partial H}{\partial x} \\ \dot{y} &= \frac{\partial H}{\partial p_y}, \quad \dot{p}_y = -\frac{\partial H}{\partial y} \end{aligned} \tag{5.22}$$

and since H does not depend on x also in this format we see that p_x is a conserved quantity, in a formula:

$$\dot{p}_x = -\frac{\partial H}{\partial x} = 0.$$

Hero–Snell revisited. We observe that both the energy and the momentum are conserved along a light ray:

$$H(x, y, p_x, p_y) = E \quad \text{and} \quad p_x = I$$

for given constants E and I. With help of E and I we easily define a third conserved quantity

$$S(x, y, \dot{x}, \dot{y}) = \frac{I}{\sqrt{2E(x, y, \dot{x}, \dot{y})}}$$

and then at once see that

$$S(x, y, \dot{x}, \dot{y}) = n(y)\frac{\dot{x}}{\sqrt{\dot{x}^2 + \dot{y}^2}} = n(y)\sin\alpha(\dot{x}, \dot{y}),$$

where α indicates the inclination of the velocity vector $\dot{q}$ with the vertical direction. Compare Figure 2.2 and the proof of Theorem 2.1. This is exactly the conservation law (2.8), found previously by more elementary means.

5.2.2 The round earth atmosphere

For the round earth atmosphere we have $q = (\varphi, r)$ and $n = n(r)$. According to (5.4) we now get

$$L(q, \dot{q}) = \frac{1}{2}\langle \dot{q}, \dot{q} \rangle_G = \frac{1}{2} n^2(r) \left(r^2 \dot{\varphi}^2 + \dot{r}^2 \right), \tag{5.23}$$

where it is taken into account that φ is an *angular* variable (measured in radians). The latter means that infinitesimal distances in the φ–direction have the form $r d\varphi = r\dot{\varphi} dt$, also compare Figure 2.8. From this we conclude that φ is indeed a cyclic variable and thus

$$p_\varphi = r^2 n^2(r)\, \dot{\varphi} \tag{5.24}$$

is conserved. It is customary to call p_φ the *angular momentum*. Further introduction of $p_r = n^2(r)\dot{r}$, see (5.9), in this case gives the Hamiltonian (5.11) the form

$$H(p, q) = \frac{1}{2n^2(r)} \left(p_\varphi^2 + p_r^2 \right). \tag{5.25}$$

The corresponding canonical equations (5.12) in this case are

$$\begin{aligned} \dot{\varphi} &= \frac{\partial H}{\partial p_\varphi}, & \dot{p}_\varphi &= -\frac{\partial H}{\partial \varphi} \\ \dot{r} &= \frac{\partial H}{\partial p_r}, & \dot{p}_r &= -\frac{\partial H}{\partial r} \end{aligned} \tag{5.26}$$

and, since H does not depend on φ, also in this way it follows that p_φ is a conserved quantity:

$$\dot{p}_\varphi = -\frac{\partial H}{\partial \varphi} = 0.$$

Hero-Snell re-revisited. For this case both

$$H(\varphi, r, p_\varphi, p_r) = E \quad \text{and} \quad p_\varphi = M$$

are conserved. And again we define a third conserved quantity

$$C(r, \varphi, \dot{r}, \dot{\varphi}) = \frac{M}{\sqrt{2E(r, \varphi, \dot{r}, \dot{\varphi})}}, \tag{5.27}$$

and easily deduce that

$$
\begin{aligned}
C(r, \varphi, \dot{r}, \dot{\varphi}) &= \frac{r^2 n^2(r)\dot{\varphi}}{n(r)\sqrt{r^2\dot{\varphi}^2 + \dot{r}^2}} \\
&= rn(r)\frac{r\dot{\varphi}}{\sqrt{r^2\dot{\varphi}^2 + \dot{r}^2}} \\
&= rn(r)\sin\alpha(\dot{r}, \dot{\varphi}),
\end{aligned}
\tag{5.28}
$$

where α is the inclination of the velocity vector $\dot{q}$ with the radial direction. The argument here is analogous to that of § 5.2.1 and we so retrieve the conservation law (2.14) derived earlier in § 2.3.

Remarks.

– Compare the conservation of angular momentum as this occurs in the case of a central force field. This conservation law corresponds to Kepler's Second Law, and it is from this case that the term 'cyclic variable' historically was derived [7, 43].

– Conservation of Energy can also be viewed in terms of symmetry and the Noether Principle, namely by observing that both the Lagrangian L and the Hamiltonian H do not explicitly depend on the parametrization by τ. The symmetry then is formed by translations in the τ–direction.

5.3 Reduction of the symmetry

The translation and rotation symmetries for the flat and the round earth allow us to reduce the dynamics from the four-dimensional (p, q)–space to a plane. Also this is analogous to the reduction of the dynamics in the central force field [7, 43]. Our starting point is formed by the formulas (5.21) and (5.25).

Theorem 5.4 (Reduction). *If the conservation laws* (5.20) *and* (5.24) *are written as*

$$p_x = I \quad \text{and} \quad p_\varphi = M, \tag{5.29}$$

for given constants I and M, then the Hamiltonians (5.21) *and* (5.25) *let themselves be reduced to*

$$\begin{aligned} H_I(y, p_y) &= \frac{1}{2n^2(y)} p_y^2 + V_I(y), \quad \textit{with} \quad V_I(y) = \frac{I^2}{2n^2(y)} \\ H_M(r, p_r) &= \frac{1}{2n^2(r)} p_r^2 + V_M(r), \quad \textit{with} \quad V_M(r) = \frac{M^2}{2r^2 n^2(r)}, \end{aligned} \tag{5.30}$$

where I and M act as parameters.

Proof. Without loss of generality we only consider the case of the flat earth. The canonical equations for this case (5.22) can be further elaborated to

$$\begin{aligned} \dot{x} &= \frac{1}{n^2(y)} p_x, \quad \dot{p}_x = 0 \\ \dot{y} &= \frac{1}{n^2(y)} p_y, \quad \dot{p}_y = \frac{n'(y)}{n^3(y)} (I^2 + p_y^2), \end{aligned} \tag{5.31}$$

where in the latter formula we applied the substitution $p_x = I$.

We see that the two equations in the second line decouple from the two in the first line, which implies that the dynamics in the (y, p_y)–direction, for $p_x = I$, can be studied separately: this is what reduction of the translation symmetry comes down to. We moreover see that the reduced equations

$$\dot{y} = \frac{1}{n^2(y)} p_y, \quad \dot{p}_y = \frac{n'(y)}{n^3(y)} (I^2 + p_y^2)$$

have the format

$$\dot{y} = \frac{\partial H_I}{\partial p_y}, \quad \dot{p}_y = -\frac{\partial H_I}{\partial y},$$

with

$$H_I(y, p_y) = \frac{1}{2n^2(y)} p_y^2 + V_I(y), \quad \text{and} \quad V_I(y) = \frac{I^2}{2n^2(y)},$$

which exactly yields the first formula of (5.30).

Reduction of the canonical system (5.26) to the second formula in (5.30) runs completely analogously. □

By using the symmetry, both systems are reduced to the (y, p_y)–, respectively the (r, p_r)–plane. We here observe that the integral curves of (5.30) are precisely the level-curves of the functions H_I or H_M, i.e., curves with equation $H_I(y, p_y) = \text{cst.}$ and $H_M(r, p_r) = \text{cst.}$ This makes it relatively simple to provide reduced phase portraits in the (y, p_y)–, respectively the (r, p_r)–plane.

The functions V_I and V_M are known as the *effective potentials* of these reductions, and are completely determined by the refraction index profiles $n(y)$ and $n(r)$. In the sequel we shall see how the associated reduced phase portraits play a role in the analysis of the optical setting.

Remarks.

– Let us perform a test first. In vacuo where $n(y) \equiv 1$, the phase portrait of the first equation in (5.30) consists of horizontal lines of the form

$$p_y = \pm\sqrt{2E - I^2},$$

where E is the value of the energy. Note that necessarily $E \geq \frac{1}{2}I^2$. Along these lines the point (y, p_y) moves with the constant speed p_y either to the left or to the right. In the reconstruction to four-dimensional space the same holds for the point $(x, p_x) = (x, I)$ that moves with velocity I to the left or to the right. In a word, we retrieve the straight lines in vacuo. Analogously for any constant n we always retrieve the familiar straight lines.

– In general the integral curves are of the form

$$p_y = \pm\sqrt{2n^2(y)E - I^2}.$$

Example 5.1 (The Poincaré upper halfplane revisited). We again consider the Poincaré upper halfplane $y > 0$ of Example 2.1. This concerns the flat earth atmosphere with refraction index profile $n(y) = 1/y$. From (5.19) and (5.21) we directly see that

$$L(x, y, \dot{x}, \dot{y}) = \frac{1}{2y^2}(\dot{x}^2 + \dot{y}^2) \quad \text{and} \quad H(x, y, p_x, p_y) = \frac{1}{2}y^2\left(p_x^2 + p_y^2\right),$$

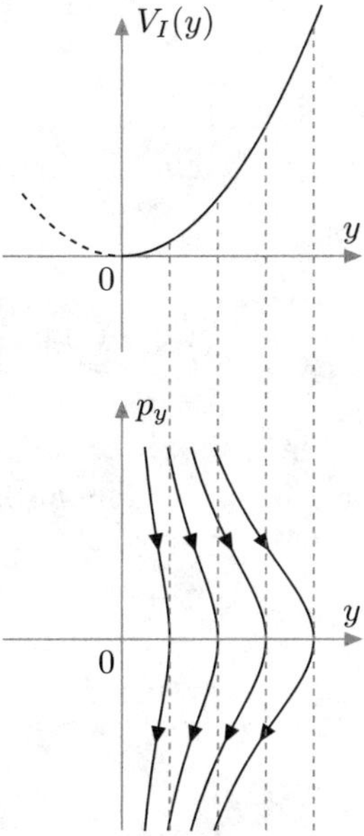

Figure 5.2. Effective potential and reduced phase portrait of the geodesic problem on the Poincaré upper halfplane. See the text for further discussion.

where the momentum (5.20)

$$p_x = \frac{1}{y^2}\dot{x}$$

is a conserved quantity. Setting $p_x = I$ as before, then according to the first formula of (5.30) we have

$$H_I(y, p_y) = \frac{1}{2}y^2 p_y^2 + V_I(y), \quad \text{with} \quad V_I(y) = \frac{1}{2}I^2 y^2.$$

See Figure 5.2 for a reduced phase portrait. Together with Figure 2.10 on page 38, this gives a geometric, qualitative impression of the dynamics in the four-dimensional (x, y, p_x, p_y)–phase space. At the end of Chapter 6 we shall return to this subject.

5.4 Scholium

As announced in the Scholium of Chapter 3, the technical tools for visualizing the information on the refraction index profiles $n = n(y)$ and $n = n(r)$, are given by the effective potentials

$$V_I(y) = \frac{I^2}{2n^2(y)} \quad \text{en} \quad V_M(r) = \frac{M^2}{2r^2n^2(r)}\,.$$

These expressions lead us from the reduced phase portraits to the sight ray geodesics. In both case the configuration space is exactly the (x, y)– or the (r, φ)–plane.

Applications. In Chapter 6 these methods will be applied to Bernoulli's brachistochrone problem, as an addition to Chapter 4. Again the refraction index profile $n(y) = 1/\sqrt{-2gy}$, $y < 0$ will play a role, originating from conservation of energy under constant gravitation with acceleration g.

In Chapter 7 we carry out the present program for the symmetric round earth atmosphere, in which case the corresponding geodesic problem can be isometrically embedded on a surface of revolution in $\mathbb{R}^3$.

Integrable systems. The geometric optics in the atmospheres of the flat and round earth leads to canonical equations that are *completely integrable* by their symmetries and the corresponding conservation laws. The central force field – as well as the associated two-body problem – and the geodesic problem on a surface of revolution are also examples of such integrable systems. It has turned out that also the geodesic problem on an arbitrary ellipsoid is integrable and in higher dimensions we find examples in the Lagrange– and Euler–top, compare [7, 43]. In all these cases (if necessary in several steps) it is possible to obtain a reduced system like (5.30), defined on a plane. This summarizes the classical strategy: look for conservation laws and reduce the dimension. It took until the end of the 19th century for Poincaré to come to the insight that a system like the three-body problem in general is non-integrable.

A system that is a small perturbation of an integrable system is often called *nearly integrable*. It is now known that 'most' canonical systems are non-integrable. A question arises what is the optical implication of this. We shall briefly return to this question in Chapter 8.

Remarks.

- We should note at once that the Hamilton Principle does not only apply to optics, but extends over almost all of classical mechanics [48, 49]. See [7, 20, 43, 62].

- The Variational Principle of Principle of Least Action extends even much further than theoretical physics, for instance see [63]. Also in extremization problems in econometrics and in the engineering sciences one resorts to such principles. A further extension of this theory is Optimal Control Theory, as developed by Pontryagin and Bellman, see Kirk [58]. For a historic perspective see Sussmann and Willems [91]. In general the context is expanded to partial differential equations.

Chapter 6

The brachistochrone revisited

We again consider Bernoulli's brachistochrone, this time armed with the technology of Chapter 5. This provides a nice illustration of the theory and on its own it also sheds new light on the brachistochrone problem. As in Chapter 4, we use the geometric optics of the flat earth atmosphere and reduce the dynamics according to Theorem 5.4, making use of the horizontal translation symmetry and the associated cyclic variable x. From the reduced problem we then reconstruct the full 4–dimensional system, which allows us to the retrieve the cycloid in a more transparant way.

This method is very similar to the one used for the motion in a central force field, where the rotationsl symmety of Kepler's Second Law leads to a 2–dimensional reduction, which then is reconstructed to the original 4–dimensional phase space. Compare [7].

6.1 Reduction of the translation symmetry

In § 4.3 we saw in detail how geometric optics is used in Bernoulli's solution of the brachistochrone problem. In fact it was the setting of the flat earth atmosphere with coordinates (x, y), where the refraction index profile $n = n(y)$ is chosen according to conservation of energy for motion under constant vertical gravitation (4.15)

$$\frac{1}{2}v^2(y) + gy = 0,$$

where g is the gravitational acceleration. For details see the discussion in Chapter 4. As before, this leads to the refraction index profile

$$n(y) = \frac{1}{\sqrt{-2gy}}, \tag{6.1}$$

where $y < 0$. This in turn leads to a reduced Hamiltonian as in (5.30), with the effective potential

$$V_I(y) = \frac{I^2}{2n^2(y)} = -I^2 gy.$$

Here I is a real number that parametrizes the reduced problem. The question then is what we can learn from the reduced phase portrait and the corresponding dynamics.

To start with we conclude that

$$\begin{aligned} H_I(y, p_y) &= \frac{1}{2n^2(y)} p_y^2 + V_I(y) \\ &= -gy\left(p_y^2 + I^2\right) \end{aligned} \tag{6.2}$$

where $p_y = -\dot{y}/(2gy)$. The reduced canonical equations (5.3) then become

$$\begin{aligned} \dot{y} &= -2gyp_y \\ \dot{p}_y &= g\left(p_y^2 + I^2\right). \end{aligned} \tag{6.3}$$

The integral curves of (6.3) are known to coincide with the level curves of H_I. Notice that for $I \neq 0$ no equilibria occur. In fact we can directly solve the equation $H_I(y, p_y) = E$ as

$$y = -\frac{1}{g}\left(\frac{E}{p_y^2 + I^2}\right), \tag{6.4}$$

see Figure 6.1. We immediately see that y as a function of p_y is minimal for

$$p_y = 0 \quad \text{with value} \quad y = -\frac{E}{gI^2}. \tag{6.5}$$

The limiting vertical line $y = 0$ in this problem has a special meaning. Although this line forms the level curve of H_I for the level $E = 0$, none of the

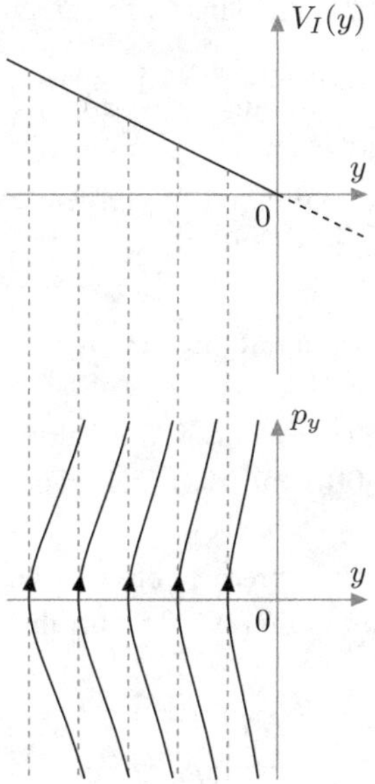

Figure 6.1. Reduced phase portrait (6.3) for Bernoulli's brachistochrone.

expressions (6.1), (6.2) and (6.3) is defined here. Yet all integral curves have the line $y = 0$ as an asymptote, both for $p_y \to +\infty$ and $p_y \to -\infty$.

A rough description of the phase portrait in Figure 6.1, on page 121, now runs as follows. Starting at $y = 0$ and $p_y = -\infty$ the y–coordinate decreases to $y = -E/(gI^2)$, after this y increases back to 0, where $p_y = +\infty$. For $(y, p_y) = (-E/(gI^2), 0)$ we have $\dot{y} = 0$, the vertical velocity vanishes.

6.2 The cycloid recovered

We explicitly compute the time parametrization of the integral curves in the reduced system (6.3). It follows from the equation $\dot{p}_y = g(p_y^2 + I^2)$ in (6.3) that

$$d\tau = \frac{dp_y}{g\big(p_y^2 + I^2\big)} = \frac{1}{gI}\frac{du}{u^2 + 1},$$

using the substitution $p_y = Iu$. This form can be integrated to

$$\tau = \frac{1}{gI}\arctan u = \frac{1}{gI}\arctan\left(\frac{p_y}{I}\right).$$

Everything now can be expressed in elementary functions. Indeed, we can write p_y in terms of τ as follows:

$$p_y = I\tan(gI\tau), \text{ which implies } y(\tau) = -\frac{E}{gI^2}\cos^2(gI\tau) \tag{6.6}$$

using (6.4). Here for $\tau = 0$ we took the value $p_y = 0$ that corresponds to the minimum (6.5) of the function (6.4). The entire motion covers the interval $|\tau| < \pi/(2gI)$.

From here we can also directly retrieve the parametrization of x in terms of τ. Indeed, from (5.29) it is known that $I = p_x = n^2(y)\dot{x}$ which implies

$$\dot{x} = \frac{1}{n^2(y)}p_x = \frac{1}{n^2(y)}I.$$

From this and (6.1) it follows that

$$\begin{aligned}\dot{x} &= \frac{1}{n^2(y)}I = -2Igy \\ &= \frac{2E}{I}\cos^2(gI\tau) = \frac{E}{I}(1+\cos(2gI\tau))\end{aligned}$$

and so

$$x = x_0 + \frac{E}{I}\left(\tau + \frac{1}{2gI}\sin(2gI\tau)\right).$$

If in (6.6) we also express y in terms of the double angle, we get

$$\begin{aligned}x &= x_0 + \frac{E}{2gI^2}(2gI\tau + \sin(2gI\tau)) \\ y &= -\frac{E}{2gI^2}(1+\cos(2gI\tau))\end{aligned} \tag{6.7}$$

and again meet the cycloid, see Figure 6.2.

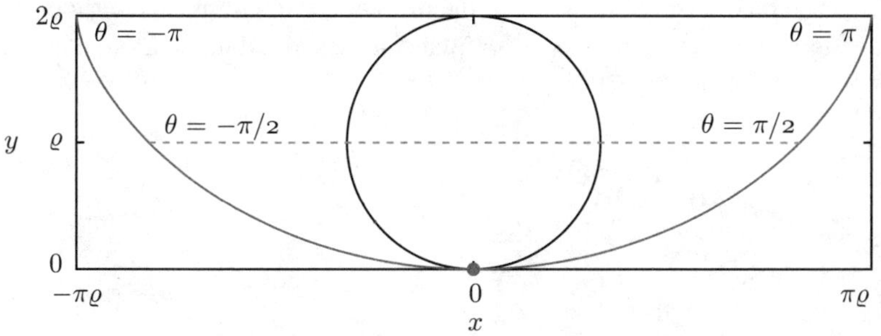

Figure 6.2. Once more we show the cycloid from Figure 4.2, now with rolling angle $\theta = 2gI\tau$ and radius $\varrho = E/(2gI^2)$. If in (6.7) we take $x_0 = -\pi\varrho = -E/(2\pi gI^2)$, the resulting cycloid is a downward translation of the above picture over $2\varrho = E/(gI^2)$.

Remarks.

– Comparing (6.7) with the earlier formulas (4.16) and (4.1) reveals that

$$\frac{E}{2gI^2} = \frac{1}{4S^2g},$$

which is in agreement with (5.2.1). Interesting moreover is the affine proportion

$$gI\tau = \frac{1}{2}\theta = \alpha - \frac{1}{2}\pi \tag{6.8}$$

between the parameters τ, θ and the inclination α with the y–direction.

– The latter is remarkable in the light of § 5.1.1, where it is asserted that light rays are geodesics of the Riemannian metric G, which by (5.5) and (6.1) has the form

$$G_q(\dot{q}, \dot{q}) = -\frac{1}{2gy}\left((\dot{x})^2 + (\dot{y})^2\right),$$

with $q = (x, y)$ and $\dot{q} = (\dot{x}, \dot{y})$.

The affine relation (6.8) therefore concerns the geodesic parameter τ and the inclination α, which means that α also forms a geodesic parameter [88].

– The reader is invited to adapt the present section to the geodesic problem on the Poincaré upper halfplane, as described in the Examples 2.1 and 5.1 from the Chapters 2 and 5.

6.3 Scholium

We conclude this chapter with two remarks.

Computations in terms of elementary functions. The above derivation of the cycloid as the brachistochrone curve is reminiscent of the way in which Newton's mathematical principles yield the Keplerian conic section orbits for the motion in a Newtonian central force field, compare [7, 43]. It seems remarkable that many of the computations of Huygens, Newton and the Bernoullis fell completely within the framework of elementary functions. Here is should be noted that already for the motion of a planar mathematical pendulum the time parametrization leads to an elliptic integral.

A 'modern' solution of the brachistochrone problem. The brachistochrone problem was one of the first in the development of variational calculus. For an overview of this extremely interesting history we refer to Goldstine [44]. In this development the Euler–Lagrange equations play a major role. Therefore we here show a simple version of the brachistochrone problem in this format. This is the way in which nowadays mathematics and physics students meet this, most often as a classroom exercise.

We briefly recall the problem from Chapter 4. The brachistochrone curve connects the point A and B in a vertical plane, in which we have a constant vertical gravitation with acceleration g. The brachistochrone curve then is a wire profile along which a bead slides without friction and such that the passage from A to B takes the shortest time.

Let the desired curve be described as a function $y = y(x)$ (curves with vertical segments can be directly excluded). As we saw in Chapter 4.2, for a piece of arc the arclength by the Pythagorean Theorem amounts to

$$ds = \sqrt{(dx)^2 + (dy)^2} = \sqrt{1 + (y')^2}\, dx, \tag{6.9}$$

where $y' = dy/dx$. From the energy conservation we again get the velocity

$$v = \frac{ds}{dt} = \sqrt{-2gy}\,, \tag{6.10}$$

see (4.3.2) or the beginning of § 6.1. From (6.9) and (6.10) it further follows that

$$dt = \sqrt{-\frac{1 + (y')^2}{2gy}}\, dx.$$

An argument as in § 5.1 of Chapter 5 now provides us with a Lagrangian

$$L(y, y') = \sqrt{-\frac{1 + (y')^2}{2gy}}. \tag{6.11}$$

The corresponding Euler–Lagrange equation (5.7) for this case reads

$$\frac{\partial L}{\partial y} - \frac{d}{dx}\frac{\partial L}{\partial y'} = 0,$$

which after after a brief calculation can be rewritten as a first order differential equation

$$\left(1 + (y')^2\right) y = \text{constant}. \tag{6.12}$$

Substitution of a parametrized solution $(x, y) = (x(\theta), y(\theta))$ then leads, as before, to the cycloidal answer, compare (4.1) and (6.7). Compare, among others, Bottema [20].

Chapter 7

Light rays as geodesics on a surface of revolution

In Chapter 5 we interpreted the light ray in an isotropic medium as a geodesic in an appropriate Riemannian metric, the dynamics of which can be described both by the Euler–Lagrange and the canonical equations (5.7) and (5.12). Just as in Chapter 3 we restrict ourselves to the round earth atmosphere with a rotation symmetric refraction index profile.

This approach makes us land in the world of rotationally symmetric differential geometry of surfaces. A very well–known type of these is formed by surfaces of revolution in 3-dimensional space, and the question addressed in this chapter is to what extent surfaces of revolution can be of interest to optics.

7.1 Summary of the chapter

Below we shall first treat the geodesic problem on a surface of revolution as such [7, 30] thereby applying the complete armament of Chapter 5. The rotational symmetry introduces a conservation law according the Noether Principle, in this case the conservation of angular momentum, compare § 5.2.2. In the same way as in (5.27), from this and the conservation of energy we obtain another conservation law, known as the Clairaut's Theorem (Alexis Claude Clairaut 1713–1765) [7, 30]. Here we apply the reduction method from § 5.3, in particular Theorem 5.4, thereby reducing the number of dimensions from 4 to 2.

After this we make the connection to optics, where necessary and sufficient conditions will be derived for isometrically embedding the corresponding Riemannian metric as the metric on an appropriate surface of revolution. If we have polar coordinates (r, φ) in the optical setting, where the refraction index profile is $n = n(r)$, then it will turn out that the modified refraction index profile $n_M(r) = rn(r)$, see (3.4), has to be identified as the distance to the rotation axis. This will enable an interpretation of the Wegener sector as described in § 3.3.2, in the set-up of surfaces of revolution.

In a scholium we digress on the differential geometry of a surface of revolution by understanding its geodesics as the orbits of a free particle on the surface. This forms an illustration of d'Alembert's Principle [7], which expresses that an appropriate normal force is being applied to keep the particle on the surface.

> In Chapter 5 we saw that geodesics minimize length. In fact geodesics are curves that (locally) form the shortest connections, like straight lines in Euclidean space. Determination of the geodesics is a minization problem that by the Variational Principle leads to Euler–Lagrange equations and to the equivalent canonical equations of Hamilton. This is a general story that also applies outside the context of optics. The present chapter follows the lines of this story for a surface of revolution that inherits its metric from the ambient 3–dimensional space and studies the ensuing geodesics. After this we consider which geometric optical models fit with this theory of surfaces of revolution.

7.2 Geodesics on a surface of revolution

We consider a surface of revolution $\mathcal{S} \subset \mathbb{R}^3$, that comes into existence by revolving a planar curve around a disjoint line in the same plane. We take cartesian coordinates (x, y, z), in which the given line is the z–axis

and assume the curve to lie in the (x, z)–plane, parametrized as

$$x = f(u), z = g(u), \tag{7.1}$$

where u runs over an open interval. Here always $f(u) > 0$ and for simplicity we also assume that

$$(f'(u))^2 + (g'(u))^2 \equiv 1, \tag{7.2}$$

which means that u is an arclength parameter. If we first introduce cylindrical coordinates (φ, r, z) by $x = r\cos\varphi$ and $y = r\sin\varphi$, then the surface $\mathcal{S}$ is parametrized by φ and u as follows:

$$x = f(u)\cos\varphi, y = f(u)\sin\varphi, z = g(u). \tag{7.3}$$

Note that at any point of $\mathcal{S}$ with coordinates (φ, u), the distance to the z–axis is given by $f(u)$. See Figure 7.1 on page 129.

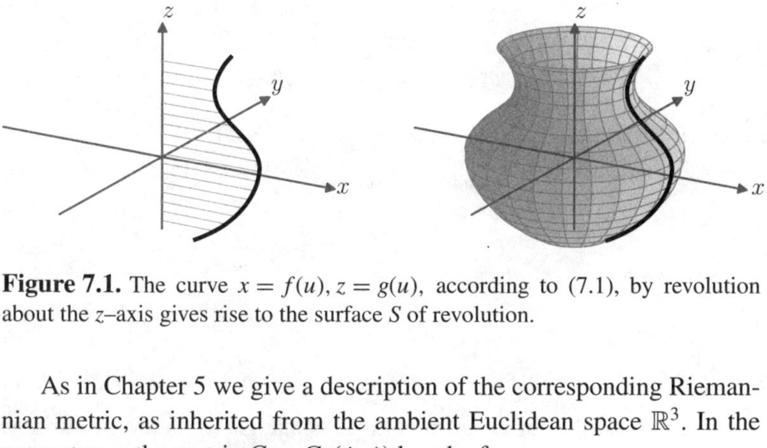

Figure 7.1. The curve $x = f(u), z = g(u)$, according to (7.1), by revolution about the z–axis gives rise to the surface S of revolution.

As in Chapter 5 we give a description of the corresponding Riemannian metric, as inherited from the ambient Euclidean space $\mathbb{R}^3$. In the present case the metric $G = G_q(\dot{q}, \dot{q})$ has the form

$$G_q(\dot{q}, \dot{q}) = \langle \dot{q}, \dot{q} \rangle_{G_q} = f^2(u)\dot{\varphi}^2 + \dot{u}^2 \tag{7.4}$$

with corresponding Lagrangian (5.23)

$$L(q, \dot{q}) = \frac{1}{2} G_q(\dot{q}, \dot{q}).$$

Here $q = (\varphi, u)$ and as in § 5.2 the factor $f^2(u)$ accounts for the fact that φ is angular. Indeed, an infinitesimal rectangle in the coordinate directions (φ, u) has edges $f(u)d\varphi$ and du. Also as before, L can be interpreted as the *kinetic energy*.

7.3 Dynamics of the geodesics

As in Chapter 5 we now describe the dynamics of the geodesics in terms of Euler–Lagrange and of Hamilton.

7.3.1 According to Euler–Lagrange

We first give a description of the geodesics in terms of the Euler–Lagrange equations (5.7).

Proposition 7.1 (Surface of revolution in Lagrangian style). *The motion along the geodesic of the surface of revolution $\mathcal{S}$ is given by the Lagrangian*

$$L(\varphi, u, \dot{\varphi}, \dot{u}) = \frac{1}{2}\left(f^2(u)\dot{\varphi}^2 + \dot{u}^2\right). \tag{7.5}$$

The corresponding Euler-Lagrange equations read

$$\begin{aligned} f^2(u)\ddot{\varphi} + 2f(u)f'(u)\,\dot{u}\,\dot{\varphi} &= 0 \\ \ddot{u} &= f(u)f'(u)\,\dot{\varphi}^2. \end{aligned} \tag{7.6}$$

The angular momentum

$$p_\varphi = f^2(u)\,\dot{\varphi} \tag{7.7}$$

is a conserved quantity.

Indeed, for the surface of revolution the angle φ is a cyclic variable. This is visible from the first equation of (7.6), which after all says nothing more than

$$\dot{p}_\varphi = 0.$$

As before we shall fix a constant M and put

$$p_\varphi = M.$$

Our proof of Proposition 7.1 consists of a few computations in the spirit of Chapters 5 and 6, the details are entrusted to the reader.

7.3.2 According to Hamilton

As in § 5.1.3 we next turn to the Hamiltonian description and so get as the equivalent counterpart of Proposition 7.1:

Proposition 7.2 (Surface of revolution in Hamiltonian style). *If* $q = (\varphi, u)$ *then* $p = (p_\varphi, p_u)$ *with*

$$p_\varphi = f^2(u)\,\dot{\varphi},\ p_u = \dot{u} \tag{7.8}$$

then the Hamiltonian is given by

$$H(p,q) = \frac{1}{2}\left(\frac{p_\varphi^2}{f^2(u)} + p_u^2\right). \tag{7.9}$$

The corresponding canonical equations read

$$\begin{aligned} \dot{\varphi} &= \frac{p_\varphi}{f^2(u)}, \quad \dot{p}_\varphi = 0 \\ \dot{u} &= p_u, \quad \dot{p}_u = \frac{p_\varphi^2 f'(u)}{f^3(u)}. \end{aligned} \tag{7.10}$$

Compare (5.25). As before $H(p,q) = L(q,\dot{q})$ is the (kinetic) energy, where $(q,\dot{q})$ and (p,q) are related according to (7.8), which is the Legendre transformation (5.9) in our situation. Also compare Theorem (5.3).

7.4 Conservation laws

As in § 5.2, the conservation laws can be written as

$$H = E \quad \text{and} \quad p_\varphi = M$$

for constants E and M. And again from these expressions we can easily derive a third one, namely that for any geodesic $\mathbf{R} = \mathbf{R}(\tau) = (\varphi(t), u(t))$ we have

$$f(u(\tau))\sin\alpha(\tau) = C. \tag{7.11}$$

Here $\alpha = \alpha(\tau)$ is the angle between the geodesic and the *meridian*; that is the 'vertical' curve $\varphi =$ constant through the point $\mathbf{R}(\tau)$. The angle α again is called *inclination*. The assertion is known as Clairaut's Theorem [7, 30].

7.4.1 Reducing the symmetry

The reduction of the symmetry now runs as in § 5.3. This means that we write

$$H_M(u, p_u) = \frac{1}{2}p_u^2 + \frac{M^2}{2f^2(u)},$$

with the effective potential

$$V_M(u) = \frac{M^2}{2f^2(u)}, \tag{7.12}$$

also compare (5.30). Here M serves as a parameter. The associated canonical equations again consist of the latter two equations of (7.10):

$$\begin{aligned} \dot{u} &= \quad p_u \\ \dot{p}_u &= -V_M'(u), \end{aligned} \tag{7.13}$$

of which the integral curves are known to be the level curves of H_M. The reduced system (7.13) exactly corresponds to the equation of motion of an oscillator

$$\ddot{u} = -V_M'(u)$$

with potential V_M. In Figure 7.2, on page 133, we present the graph of V_M near a minimum and a maximum. The corresponding reduced phase portrait then, respectively, is a *center*, or a *saddle point*.

Remarks.

- The oscillator equation of motion also is already visible as the second equation in the system (7.6). Further note that the first equation of (7.6) exactly gives the conservation law $p_\varphi = M$.

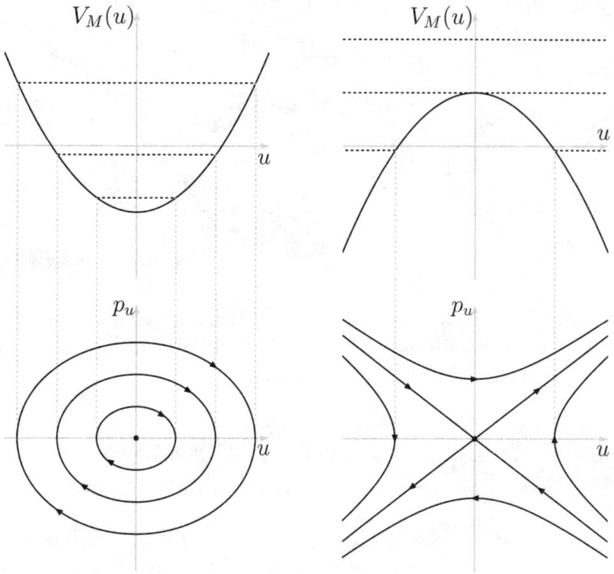

Figure 7.2. Reduced phase portraits near a minimum (left) and a maximum (right) of the effective potential $V_M = M^2/(2f^2(u))$, see (7.12).

– Consider the entire system (7.10) in the 4–dimensional $(\varphi, u, p_\varphi, p_u)$–space: the phase space of this system. Each integral curve of this projects on the 2–dimensional surface $\mathcal{S} = \{\varphi, u\}$ as a geodesic. A different projection is that on the reduced phase plane $\{u, p_u\}$, as an integral curve of the reduced system (7.13). By both 2–dimensional projections we get an insight in the 4–dimensional dynamics.

7.4.2 Description of the geodesics

Analogous to Theorem 2.5 of Chapter 2, the Clairaut Theorem (7.11) completely determines the form of the geodesic on $\mathcal{S}$ by specifying the values of M or of C. Here it is understood that always $f(u)$ indicates the distance of a point $(\varphi, u) \in \mathcal{S}$ to the z–axis. From this we can infer the following.

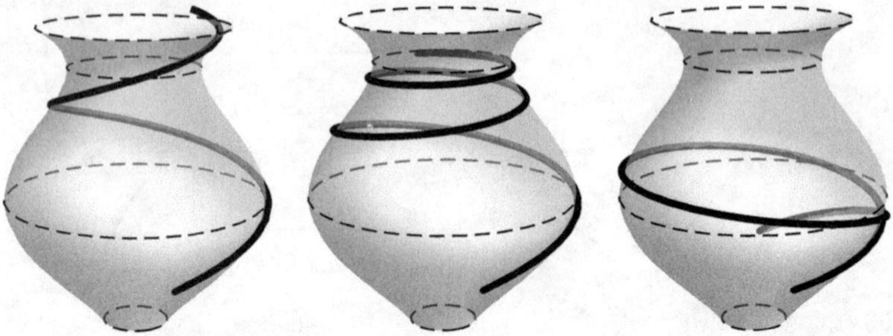

Figure 7.3. The Bottle: a surface of revolution, the geodesics of which correspond to a sight ray bundle as in Figure 3.18. Left: refraction. Center: limit angle. Right: reflection.

Lemma 7.3 (Meridians and parallels). *All meridians* φ = constant *are geodesics on* $\mathcal{S}$. *A parallel* $u = u_0$ *is a geodesic if and only if* $f'(u_0) = 0$.

Proof. From Clairaut's conservation law (7.11), which here means that

$$f(u(\tau))\sin(\alpha(\tau)) = C,$$

it immediately follows that all meridians are geodesics, indeed this corresponds to the case where

$$\sin\alpha(\tau) \equiv 0$$

and we just take $M = C = 0$.

For the statement on parallels we consider the second equation of (7.6). Here $u \equiv u_0$ which also means that $\ddot{u} = 0$, which in turn implies that

$$f(u_0)f'(u_0)\dot{\varphi}^2(\tau) = 0.$$

Our assertion follows since we always have that $f(u_0) > 0$. □

7.4.2.1 Near a geodesic parallel

With the help of the Clairaut formula (7.11) we can already get a direct insight into the qualitative organization of the geodesics, as we shall illustrate with the help of Figure 7.3 on page 134. We can clarify this description further by the reduced system (7.13) and Figure 7.2. Indeed, taking a point on $\mathcal{S}$ with coordinates (φ, u) we consider the geodesic γ on $\mathcal{S}$ that starts at this point with an inclination α. For simplicity we assume that γ is parametrized according to arclength.

In the 4–dimensional phase space, where the entire canonical system (7.10) lives, we also have to deal with the coordinates (p_φ, p_u) and consider the projection of γ on the reduced (u, p_u)–phase plane, as sketched in Figure 7.2.

We first note that a minimum of V_M corresponds to a maximum of f at the same point. Similarly a maximum in V_M corresponds to a minimum in f. Non-degeneracy of the extremes in V_M also corresponds to non-degeneracy of the corresponding extremes in f, where we use that always $f(u) > 0$.

Geodesics near a maximum $f(u_0)$. As we already saw, a maximum $f(u_0)$ corresponds to a geodesic parallel and any nearby geodesic oscillates around this parallel. This can be seen by the Clairaut formula (7.11): assuming that $C > 0$, the expression

$$f(u(\tau)) \sin \alpha(\tau) = C$$

can only hold along the geodesic if for decreasing $f(u(\tau))$, the function $\sin \alpha(\tau)$ increases. The latter has a maximum value 1, where we have that $\alpha(\tau) = \frac{1}{2}\pi$: the geodesic then is perpendicular to the meridian and is tangent to a parallel.[1]

After this $\alpha(\tau)$ decreases again and $f(u(\tau))$ increases while the geodesic returns to the widest parallel $u = u_0$, intersects this (transversally), on the other side more or less repeating this process. Compare Figure 7.3, right.

For $C < 0$ the same story occurs, with a few evident adaptations. In Figure 7.2, left, the motions correspond to oscillations around the elliptic equilibrium $(u, p_u) = (u_0, 0)$.

[1]This parallel itself cannot be a geodesic! Indeed, geodesics that are tangent at one point have to coincide because of the uniqueness of solutions of ordinary differential equations [34, 53].

Geodesics near a minimum $f(u_0)$. A minimum $f(u_0)$ also corresponds to a geodesic parallel, which is the tightest in its neighborhood. Now there are several possibilities that are most easily characterized with the help of Figure 7.2, left. Generally speaking the reduced integral curve remains in one of the four quadrants in the neighborhood of a saddle point $(u, p_u) = (u_0, 0)$ and roughly speaking it enters along one of the two asymptotes and leaves along the other one. One might say that the integral curve changes its asymptote. We note that the motion is slower when closer to the saddle point.

- **Refraction.** In the case of refraction the geodesic traverses the geodesic parallel $u = u_0$, see Figure 7.2, right and Figure 7.3, left. This simply means that the corresponding integral curve of the reduced system stays in one of the quadrants, passing above or beneath the saddle point $(u, p_u) = (u_0, 0)$, meanwhile intersecting the level $u = u_0$.
  A corresponding geodesic in the (φ, u)–plane intersects the tight geodesic with a non-zero angle, meanwhile moving somewhat in the φ–direction. In fact, the decrease of the distance $f(u(\tau))$ to the z–axis toward the minimum has to be compensated by an increase of the inclination $\alpha(\tau)$. There will be enough space left to have an inclination $|\alpha| < \frac{1}{2}\pi$ at the level $u = u_0$. See again Figure 7.3, left.

- **Reflection.** In the case of reflection we are situated left or right of the reduced saddle point: the integral curve stays in one of the areas $u < u_0$ or $u > u_0$. Compare Figures 7.2, left and Figure 7.3, right. A corresponding geodesic resides at a certain distance from the z–axis, in such a way that a further decrease of this to the value $f(u_0)$ is not allowed; this because the compensating increase of $\sin\alpha(\tau)$ has already reached the maximum value 1. In other words, the geodesic is tangent to a parallel and is reflected in the same u–direction it came from.

- **Limit angle behavior.** In the transition between refraction and reflection we have the limit angle behavior: the integral curve in the (u, p_u)–plane precisely follows a stable manifold of the saddle point $(u_0, 0)$. We note that this curve will not attain this saddle point in a finite time.

The corresponding geodesic spirals towards the tight geodesic parallel. Compare Figures 7.3, center, and 2.9.

We now will study the consequence of the analysis for the complete motion of the geodesic in the $(\varphi, u, p_\varphi, p_u)$–phase space.

7.4.2.2 Reconstruction to four dimensions

From the above we can reconstruct the dynamics of the original, entire system (7.10) in the 4–dimensional $(\varphi, u, p_\varphi, p_u)$–phase space, at least in a qualitative sense. Here we take a fixed value of the angular momentum M. The invariant hypersurface $p_\varphi = M$ is 3–dimensional and allows for a parametrization by (φ, u, p_u) and therefore is the product of the reduced (u, p_u)–phase plane and a circle parametrized by φ. If we then vary the parameter M, the 4–dimensional space is locally *foliated* by the p_φ–levels [30, 88].

To start with the equilibrium $(u_0, 0)$ of the reduced system for any value of M for the entire system yields a closed integral curve, i.e., a periodic solution.

Periodic solution of saddle type. In the case where the reduced system (7.13) at $(u_0, 0)$ has a saddle point, the 3–dimensional restriction of system (7.10) to the invariant hypersurface $p_\varphi = M$ has a closed integral curve of saddle type, see Figure 7.4, right.

Periodic solution of elliptic type. In the case where the reduced equilibrium $(u_0, 0)$ is a center, a neighborhood of the closed integral curve in the 3–dimensional hypersurface is foliated by invariant 2–dimensional tori. The dynamics on each of these tori is *multi-periodic*. This means that on such a torus either all integral curves are closed of the same period, or each integral curve is dense in this torus. Compare Figure 7.4, above and left. For background see [7, 26].

7.4.3 More global behavior of geodesics

We now consider more global behavior of geodesics as they wind themselves along a surface of revolution.

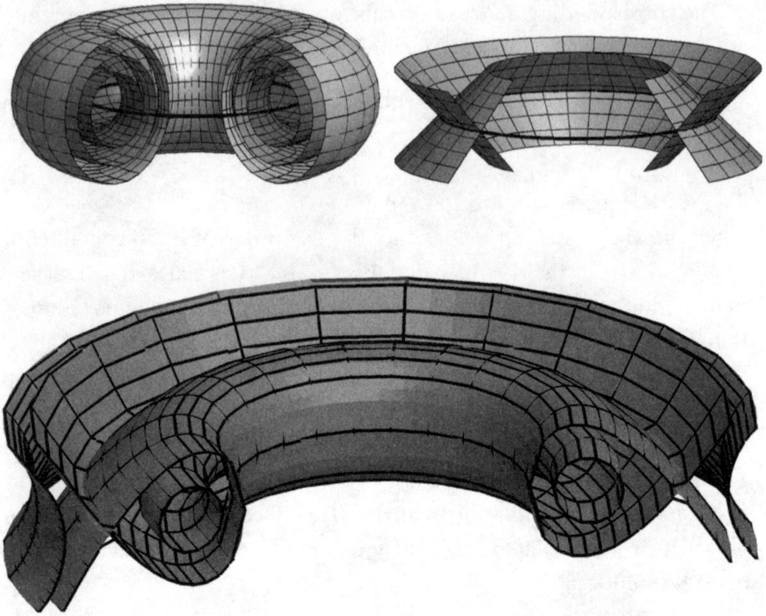

Figure 7.4. Sketch of the invariant surfaces in the dynamics of the entire system (7.10) in an invariant 3–dimensional hypersurface $p_\varphi = M$. The closed integral curve can be of saddle type (top, right) or is of elliptic type and then is surrounded by a *foliation* of invariant 2–dimensional tori (top, left). The bottom figure gives a similar reconstruction for the Bottle of Figure 7.3.

Example 7.1 (The 2–dimensional sphere). In this example we consider the 2–sphere $\mathcal{S} = \mathbb{S}^2 \subset \mathbb{R}^3$ as a surface of revolution. To be precise, we take the circle $x^2 + (z-2)^2 = 1$, parametrised as

$$f(u) = \cos(2\pi u) \quad \text{and} \quad g(u) = 2 + \sin(2\pi u),$$

$-\frac{1}{4} < u < \frac{1}{4}$. In that case for a given constant M we have for the effective potential

$$V_M(u) = \frac{M^2}{2\cos^2(2\pi u)}.$$

The reader is encouraged to sketch the surface, effective potential and the corresponding reduced phase portrait. What does the Clairaut Theorem (7.11) imply here?

All integral curves now are closed and we find ourselves in the left half of the Figures 7.2 and 7.4 (top). In this case all multi-periodic integral curves are really periodic, as can be shown by a computation. This fits with the generally known fact that the geodesics on $\mathcal{S} = \mathbb{S}^2$ precisely are the great circles, so with center $(x, y, z) = (0, 0, 2)$.

Example 7.2 (The Bottle). We present an example that is extremely relevant for our geometric-optical application, and that fits well with the theory of Chapters 2 and 3. In the following sections we shall further clarify this connection. We return to Example 3.1 of Chapter 3, also see the corresponding Figure 3.18. This concerns the simplest case of a Wegener sector with a smooth refraction index profile $n = n(r)$. In terms of a surface of revolution this nicely corresponds with the geodesics on the Bottle, as sketched in Figure 7.3.

In Figure 7.5 (page 140) we present a corresponding effective potential with a reduced phase portrait, that contains both a center and a saddle point. If the observer W is at a height $r = r_W$, then he observes a Wegener sector, namely in all observation directions that lead to integral curves contained within the loop of the saddle point.[2] In the reconstruction to the 3–dimensional (φ, u, p_u)–space, the entire figure has to be revolved around the axis $r = 0$, compare Figure 7.4 (bottom).

7.5 Geodesics and light rays

We now shall clarify the connection between the geodesics on a surface of revolution $\mathcal{S}$ and the light rays in the round earth atmosphere as described in § 5.1. From (5.23) we recall the Riemannian metric $G^{\text{opt}}_{\varphi,r}$ of the associated geometric optics as

$$G^{\text{opt}}_{\varphi,r}((\dot\varphi, \dot r), (\dot\varphi, \dot r)) = n^2(r)(r^2\dot\varphi^2 + \dot r^2).$$

[2] Often indicated as a *homoclinic* loop.

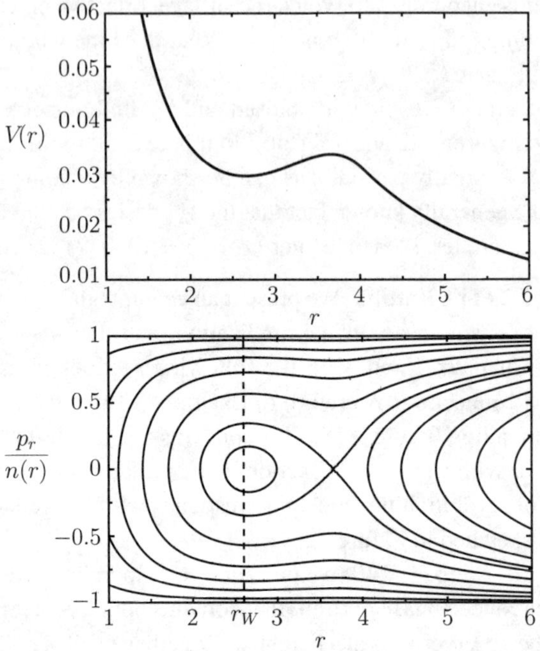

Figure 7.5. Effective potential and corresponding reduced phase portrait concerning the Bottle of Example 7.2 and Figure 7.3. As mentioned earlier, this setup fits with the light ray bundle from Figure 3.18, where we also show a modified refraction index profile.

Comparison with the metric (7.4) shows us that if we want to identify the two, the curve (7.1) needs to have the general format

$$x = f(r), z = g(r)\,, \tag{7.14}$$

i.e., parametrized by the variable r, where we will have to specify the functions f and g, which will be done in the next section. This is why, instead of (φ, u), from now on we have to use (φ, r). In this setting the Riemannian metric (7.4) of the surface of revolution gets the form

$$G^{\text{omw}}_{\varphi,r}((\dot{\varphi}, \dot{r}), (\dot{\varphi}, \dot{r})) = f^2(r)\dot{\varphi}^2 + \left((f')^2(r) + (g')^2(r)\right)\dot{r}^2 \tag{7.15}$$

and thus the Lagrangian $L^{\mathrm{omw}}(q, \dot{q}) = \frac{1}{2}G^{\mathrm{omw}}(\dot{q}, \dot{q})$ from (7.5) now is given by

$$L^{\mathrm{omw}}(\varphi, r, \dot{\varphi}, \dot{r}) = \frac{1}{2} f^2(r)\,\dot{\varphi}^2 + \frac{1}{2}\left((f')^2(r) + (g')^2(r)\right)\dot{r}^2.$$

The corresponding Euler–Lagrange equations therefore read

$$\ddot{\varphi} + \frac{2f'(r)}{f(r)}\,\dot{r}\,\dot{\varphi} = 0$$

$$\ddot{r} - \frac{1}{(f')^2(r) + (g')^2(r)}\left(f(r)f'(r)\,\dot{\varphi}^2 - (f'(r)f''(r) + g'(r)g''(r))\,\dot{r}^2\right) = 0,$$

compare (7.6). The former of these equation again expresses the conservation of angular momentum; in the $(\varphi, r, p_\varphi, p_r)$–coordinates after all we have

$$\dot{p}_\varphi = 0 \quad \text{where} \quad p_\varphi = f^2(r)\dot{\varphi}\,,$$

and we can write as before

$$p_\varphi = M$$

for a constant M, compare (7.7).

Remarks.

- This formulation in terms of $(\varphi, r, \dot{\varphi}, \dot{r})$ is equivalent to the previous one in terms of $(\varphi, u, \dot{\varphi}, \dot{u})$, we just use other coordinates. We here note the important fact that the Variational Principle, as formulated in Chapter 5, is coordinate free.
- The relationship between the arclength parameter u and the coordinate r is determined by the equation

  $$du^2 = \left((f')^2(r) + (g')^2(r)\right) dr^2,$$

  in which the Pythagorean Theorem is manifest.

7.6 Isometric embeddability

The question that we have to address is under what conditions the geometric optics of the round earth atmosphere can be considered as the geodesic problem on a surface of revolution. On the one hand we have the (φ, r)–configuration plane with Riemannian metric G^{opt} determined by the refraction index profile $n = n(r)$, and on the other side the surface of revolution, parametrized by (φ, r), with Riemannian metric G^{omw} determined by the functions $f = f(r)$ and $g = g(r)$.

So, given $n = n(r)$ we have to find $f = f(r)$ and $g = g(r)$, compare (7.14), such that the following equations are satisfied:

$$\begin{aligned} f^2(r) &= r^2 n^2(r) \\ (f')^2(r) + (g')^2(r) &= n^2(r). \end{aligned} \tag{7.16}$$

The former of these equations exactly determines f as the *modified refraction index profile* $f(r) = rn(r) = n_M(r)$, compare § 3.4, and the remaining question then is whether a function g can be found that satisfies the latter equation.

Theorem 7.4 (Isometric embeddability as a surface of revolution). *The Riemannian manifold $\mathbb{R}^2 \setminus \{0\} = \{\varphi, r\}$ with metric G^{opt} determined by the refraction index profile $n = n(r)$, see* (5.23), *can be embedded isometrically as a surface of revolution in $\mathbb{R}^3$*

$$x = f(r)\cos\varphi,\ y = f(r)\sin\varphi,\ z = g(r),$$

if and only if for each r one has

$$\frac{-2n(r)}{r} \le n'(r) \le 0. \tag{7.17}$$

Proof. According to (7.16), given the choice $f(r) = rn(r)$, also

$$(f')^2(r) + (g')^2(r) = n^2(r) \tag{7.18}$$

has to be solved for g. From the Leibniz formula it follows that $f'(r) = n(r) + rn'(r)$. Hence (7.18) becomes

$$n^2(r) + 2rn(r)n'(r) + r^2(n')^2(r) + (g')^2(r) = n^2(r),$$

or

$$(g')^2(r) = -\left(2rn(r)n'(r) + r^2(n')^2(r)\right).$$

The last equation can be solved if and only if

$$2n(r)n'(r) + r(n')^2(r) \leq 0,$$

or, equivalently, if

$$n'(r)\left(2n(r) + rn'(r)\right) \leq 0.$$

In principle we have either

$$n'(r) \leq 0 \quad \text{and} \quad 2n(r) + rn'(r) \geq 0$$

or

$$n'(r) \geq 0 \quad \text{and} \quad 2n(r) + rn'(r) \leq 0.$$

A priori we know that $n(r) = f(r)/r > 0$, which cancels the latter possibility. So we conclude that only

$$n'(r) \leq 0 \quad \text{and} \quad 2n(r) + rn'(r) \geq 0$$

remains, or equivalently,

$$\frac{-2n(r)}{r} \leq n'(r) \leq 0. \qquad \square$$

Remarks.

– The second formula of (7.17) is a differential inequality, expressing that for a certain constant $c > 0$ for all $r > 0$ necessarily

$$n(r) \geq \frac{c}{r^2}.$$

This is a strong condition when $r \downarrow 0$, but since r denotes the distance to the center of the earth, this does not have to bother us.

– A physical condition further says that $n(r) \geq 1$, which implies that for the relevant cases also $f(r) = n_M(r) \geq r$.

7.7 Scholium

As we saw in Theorem 7.4, a rotationally symmetric geometric optics, as described in Chapters 3 and 5, might or might not fit with the geodesic dynamics of a surface of revolution. We note only a few consequences, leaving a broader discussion to Chapter 8. After this we present a characterisation of the geodesics on a surface of revolution in terms of the motion of a free particle.

7.7.1 Rotation symmetry with or without surface of revolution

If we consider the conditions of Theorem 7.4, the following catches the eye. The condition for a refraction index profile $n = n(r)$

$$\frac{-2n(r)}{r} < n'(r) < 0 \tag{7.19}$$

that is necessary and sufficient for embeddability, is *open* in the C^1–topology [26, 52]. This means that small perturbations of n whose first derivative is also small still give rise to isometric embeddability. In this sense the condition (7.19) yields robust models with regards to embeddability.

Although the robust condition (7.19) covers a number of important cases of Chapter 3 and Chapter 5, we know from §§ 3.2 and 3.2.3.1 that cases exist where the conditions of Theorem 7.4 are not satisfied. In particular this holds for the Takens refraction index profiles III and IV in Table 2 of §3.2.3.1; also compare the associated profiles in Figure 3.14. In these cases the refraction index profile is not monotonically decreasing, since *lower reflection* occurs at or near the surface of the earth. It should be noted that a refraction index profile $n = n(r)$ with a non-degenerate local maximum gives rise to a model in which *non-embeddability* is a robust property.

7.7.2 Free particle on a surface of revolution

We briefly indicate how the geodesic problem on a surface of revolution can be regarded in terms of the orbits of a *free particle* moving on that surface. The adjective 'free' indicates that no gravitation or other external forces are at work, but only the force that keeps the particle on the surface. This forms a nice application of the d'Alembert Principle (Jean le Rond d'Alembert 1717–1783), which asserts that the particle is kept on the surface by a force that is perpendicular to the surface (see [7]).

Remarks.

- In terms of the pendulum, one may think of a point mass that is kept on a circle by a stiff massless rod attached to the center. Also here the force is perpendicular to the circle.
- Such a force $\mathbf{F}$ does not perform any work, since $\langle \mathbf{F}, \dot{q} \rangle \equiv 0$ for any path $\tau \mapsto q(\tau)$, compare [7, 43].

Let r, φ and z be cylindrical coordinates on

$$\mathbb{R}^3 \setminus z-\text{axis} = \{(x, y, z) \in \mathbb{R}^3, \text{ with } (x, y) \neq (0, 0)\}.$$

We have $x = r\cos\varphi$ and $y = r\sin\varphi$. In the (x, z)–plane a parametrized curve (7.1)

$$x = f(u), z = g(u)$$

is given, where u varies over an open interval and where we assume that always $f(u) > 0$.

Without loss of generality we again assume that

$$(f'(u))^2 + (g'(u))^2 \equiv 1,$$

compare (7.2), which again expresses that u is an arclength parameter. This curve is revolved around the z–axis, leading to a surface $\mathcal{S}$

$$x = f(u)\cos\varphi, y = f(u)\sin\varphi, z = g(u),$$

parametrized by φ and u, compare (7.3). We now elaborate on the dynamics of the free particle in terms of φ and u. This description is intrinsic on the surface $\mathcal{S}$, which allows us to make a connection with earlier parts of this chapter.

Let $\tau \in \mathbb{R} \mapsto \mathbf{R}(\tau) \in \mathcal{S}$ be the orbit of a free particle, then the d'Alembert Principle says that for all τ we need $\ddot{\mathbf{R}}(\tau)$ to be perpendicular to the surface $\mathcal{S}$.

This means that both $\ddot{\mathbf{R}} \perp \mathbf{e}_\varphi$ and $\ddot{\mathbf{R}} \perp f'(u)\,\mathbf{e}_r + g'(u)\,\mathbf{e}_z$ and therefore, in terms of inner products,

$$\langle \ddot{\mathbf{R}}, f'(u)\,\mathbf{e}_r + g'(u)\,\mathbf{e}_z \rangle = 0 \quad \text{and} \quad \langle \ddot{\mathbf{R}}, \mathbf{e}_\varphi \rangle = 0. \tag{7.20}$$

Here f' and g' are the derivatives of f and g with respect to u. Moreover, $\dot{\varphi} = d\varphi/d\tau$, $\ddot{r} = d^2r/d\tau^2$, etc.

From now on writing $r(\tau) = f(u(\tau))$ and $z = g(u(\tau))$, after a few calculations it turns out that (7.20) is equivalent to

$$\begin{aligned} f^2(u)\,\ddot{\varphi} + 2f(u)f'(u)\dot{u}\,\dot{\varphi} &= 0 \\ \ddot{u} - f(u)f'(u)\,\dot{\varphi}^2 &= 0\,, \end{aligned}$$

which exactly amounts to the system of Euler-Lagrange equations (7.6). This leads to familiar ground: we have proved that both approaches are equivalent. We note that the equation $\langle \ddot{\mathbf{R}}, \mathbf{e}_\varphi \rangle = 0$ is equivalent to $f^2(u)\,\ddot{\varphi} + 2f(u)f'(u)\dot{u}\,\dot{\varphi} = 0$, which again expresses the conservation of angular momentum.

Remarks.

– In the above calculations it is useful to know that the following formulas hold true for velocity and acceleration of $\mathbf{R}$:

$$\begin{aligned} \dot{\mathbf{R}} &= \dot{r}\,\mathbf{e}_r + r\dot{\varphi}\,\mathbf{e}_\varphi + \dot{z}\,\mathbf{e}_z \\ \ddot{\mathbf{R}} &= (\ddot{r} - r\dot{\varphi}^2)\,\mathbf{e}_r + (2\dot{r}\dot{\varphi} + r\ddot{\varphi})\,\mathbf{e}_\varphi + \ddot{z}\,\mathbf{e}_z, \end{aligned} \tag{7.21}$$

where $\mathbf{e}_r$, $\mathbf{e}_\varphi$ and $\mathbf{e}_z$ are the unit vectors in the r–, the φ– and the z–direction.

– The d'Alembert Principle embodies, next to the length-minimizing Euler–Lagrange equations, an equivalent characterization of geodesics, which amounts to the requirement that the covariant derivative of $\ddot{\mathbf{R}}$ equals 0, which in turn means that γ is a geodesic if $\dot{\gamma}$ is a parallel vector field along γ. Compare [30, 88].

– Concerning the general problem of isometric embeddability we would like to mention the famous Theorem of Nash–Moser that asserts that any Riemannian manifold can be embedded isometrically in $\mathbb{R}^N$ for sufficiently large N. In Theorem 7.4 evidently we are dealing with $N = 3$. Again see [30, 88].

Chapter 8
Scholium generale

Beyond the Chapters 2, 3, 5 and 7, and in particular § 7.7, there is a lot more to say about optics, even geometric optics. There is ample intriguing photographic material, compare the Plates IIIa and IIIb and Figure 62 in Minnaert [72], the latter reproduced here as Figure 1.6 on page 8. Also compare Van der Werf et al. [95, 96] and Floor [39–41], but the internet easily provides the curious reader with a wealth of further examples. We shall not try to compete with all this here, but restrict ourselves to giving a few more theoretical impressions.

A first comment concerns acoustics: indeed the entire theory of this book also applies when 'light' is replaced by 'sound' [79]. One interesting complication in that case is that the medium, the air, may change considerably with time.

8.1 Approximate rotational symmetry

In Chapter 5, particularly in § 5.1, we discussed the eikonal equations for light rays in the 3–dimensional atmosphere. Here we we would like to consider the optics on a vertical plane under the assumption that the refraction index is *nearly* symmetric, i.e., we consider a small perturbation of the rotationally symmetric case. In particular, consider Examples 3.1 and 7.2 concerning the Bottle, and assume that the perturbation is periodic in the angle variable φ as follows:

$$n(\varphi, r) = n_0(r) + \varepsilon n_{\mathrm{pert}}(\varphi, r). \tag{8.1}$$

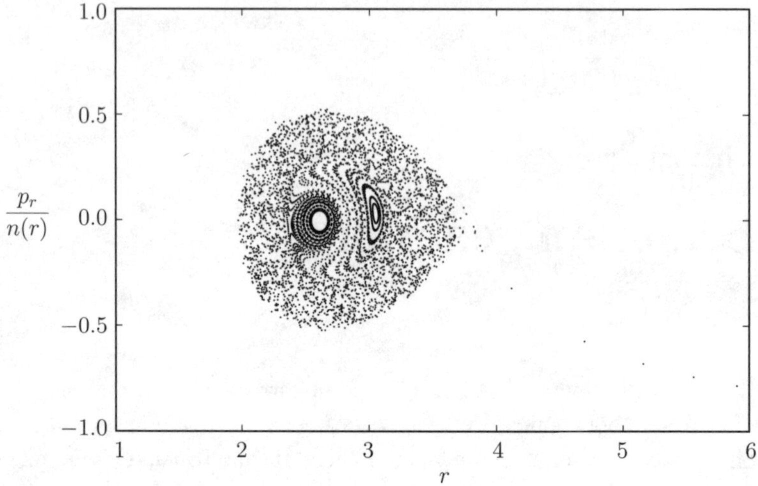

Figure 8.1. Iso-energetic Poincaré map of the periodically perturbed Bottle, compare Figure 7.5. The orbits of this map have come into existence by iterations that follow the integral curves of the canonical equations (5.26) while they go around the earth. As in many nonlinear systems, we see the co-existence of order and chaos.

Here think of $|\varepsilon|$ as a small real number. For $\varepsilon = 0$ we encounter the unperturbed, rotationally symmetric refraction index profile, The unperturbed, rotationally symmetric refraction index profile as this occurs for $\varepsilon = 0$, gives rise to a reduction to the (r, p_r)–plane, where $p_r = n_0^2(r)\dot{r}$, compare (5.9) and § 5.2.2. In Figure 7.5 on page 140, we presented a reduced phase portrait.

For $\varepsilon \neq 0$ the rotational symmetry disappears and therefore the conservation of the angular momentum p_φ, see (5.24), is no longer true. On the other hand, the energy

$$H(\varphi, r, p_\varphi, p_r) = \frac{1}{2n^2(\varphi, r)}\left(p_\varphi^2 + p_r^2\right),$$

see (5.11), remains a conserved quantity. In such circumstances, within a fixed energy level $H(\varphi, r, p_\varphi, p_r) = E$ an *iso-energetic Poincaré map* can be defined that maps a given (r, p_r)–plane to itself, compare [26]. Here it is important to note that in the 3–dimensional energy level, the integral curves of the canonical equations (5.26) intersect such a plane transversally. The Poincaré map follows an integral curve of the canonical system (5.26) while this goes around the earth and the orbit then is formed by the succesive intersections with this plane, see Figure 8.1.

A global interpretation of this is that the dynamics partly is determined by the orbits that constitute invariant circles and partly by the chaotic clouds [26]. The invariant circles in the 4–dimensional phase space correspond to invariant 2–dimensional tori as these also show up in Figure 7.4.

Remarks.

- The reader may well ask what is the physical significance of Figure 8.1, particularly since in the lower atmosphere the light rays are extinguished soon. What we hope to illustrate is that the problem in this form already has the full complexity of generic dynamics, that is, within the world of planar area preserving maps.
- In more realistic circumstances the refraction index profile will moreover be time dependent, and even more so in the case of acoustics. To make things even worse we generally will not be allowed to restrict ourselves to one vertical plane, as an example again see Figure 1.6 on page 8.
- The difference between $n_0(r)$ and $n(\varphi, r)$ in (8.1) is that between an integrable and a nearly integrable system, compare Scholium 5.4. This difference historically has been of great importance for the development of the theory of nonlinear dynamical systems,[1] for more information see [7, 12, 26].

[1] ... nowadays often known as 'chaos theory'.

8.2 A connection with wave optics

The Fermat Principle follows mathematically from the Huygens Principle, which asserts that light progresses in wave fronts, where every point of a wave front is the center of a new wave front. For an illustration of Snell's Law within this theory see Figure 8.2 on page 153. The light rays here act as the normals on the wave fronts. For more details see Arnold [7], § 46 or Feynman et al. [38] part I, § 26. For later optical wave theories, in particular the electromagnetic theory of Maxwell, we refer to [46, 81]. A nice description of quantum electrodynamics is given by Feynman [37]. In Chapter 5 in Part II, the Fermat Principle was turned into Hamilton's Principle and the Variational Principle, thereby considerably deepening our mathematical insights in geometric optics. Also the theories of wave theoretic optics can be founded on the Variational Principle, as in the above references.

8.3 Caustics

A *caustic* arises when light rays accumulate on each other. In Figure 8.3 (on page 154), left, we depict a cusp as this occurs as the envelope of a ray bundle generated by a circular light source. For another example compare Figure 4.11. A cuspidal caustic, under suitable incidence of light, can already be observed on the bottom of an almost empty coffee cup.

In Catastrophe Theory [8, 81], and more generally in Singularity Theory [9, 11, 46], geometric shapes of caustics are analyzed, as these can occur in many different appearances and that moreover have a certain robustness. Of the cusp we already saw a few examples. Also the *fold* is a famous catastrophe, that among others occurs in the rainbow, where it can be observed in a frequency dependent form [72, 81].

It is interesting to study refined versions of caustics within the theories of wave optics. For the cusp see Figure 8.3, center and right; compare Poston and Stewart [81] p. 266. For beautiful descriptions and

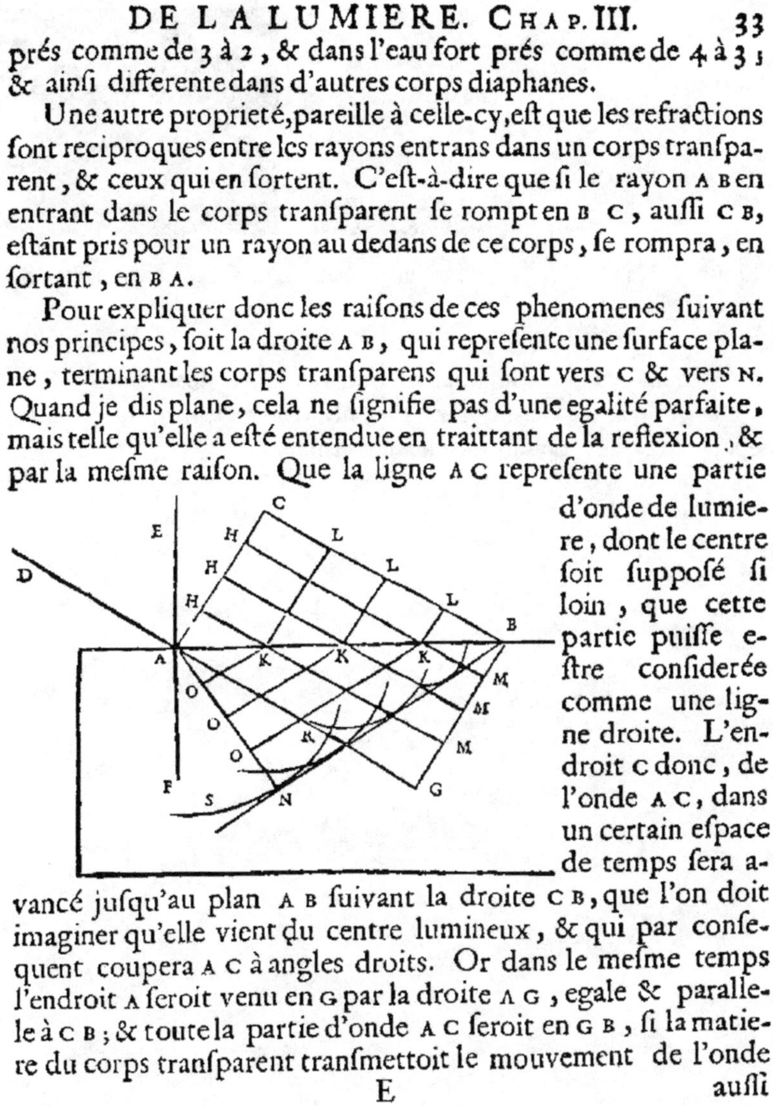

DE LA LUMIERE. CHAP. III. 33

prés comme de 3 à 2, & dans l'eau fort prés comme de 4 à 3; & ainſi differente dans d'autres corps diaphanes.

Une autre proprieté, pareille à celle-cy, eſt que les refractions ſont reciproques entre les rayons entrans dans un corps tranſparent, & ceux qui en ſortent. C'eſt-à-dire que ſi le rayon A B en entrant dans le corps tranſparent ſe rompt en B C, auſſi C B, eſtant pris pour un rayon au dedans de ce corps, ſe rompra, en ſortant, en B A.

Pour expliquer donc les raiſons de ces phenomenes ſuivant nos principes, ſoit la droite A B, qui repreſente une ſurface plane, terminant les corps tranſparens qui ſont vers C & vers N. Quand je dis plane, cela ne ſignifie pas d'une egalité parfaite, mais telle qu'elle a eſté entendue en traittant de la reflexion, & par la meſme raiſon. Que la ligne A C repreſente une partie d'onde de lumiere, dont le centre ſoit ſuppoſé ſi loin, que cette partie puiſſe eſtre conſiderée comme une ligne droite. L'endroit C donc, de l'onde A C, dans un certain eſpace de temps ſera avancé juſqu'au plan A B ſuivant la droite C B, que l'on doit imaginer qu'elle vient du centre lumineux, & qui par conſequent coupera A C à angles droits. Or dans le meſme temps l'endroit A ſeroit venu en G par la droite A G, egale & parallele à C B; & toute la partie d'onde A C ſeroit en G B, ſi la matiere du corps tranſparent tranſmettoit le mouvement de l'onde

E auſſi

Figure 8.2. Snell's law with the wave optics of Huygens [56].

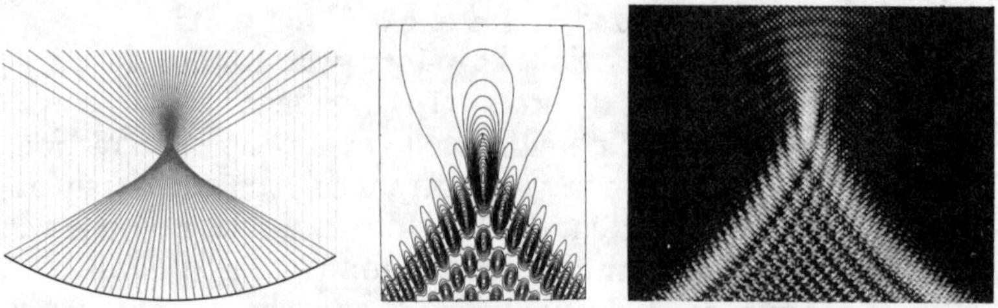

Figure 8.3. Left: Cuspidal caustic in geometric optics. Center and right: Wave optical refinements [81] p. 266.

background also see Sir Michael Berry [15, 16]. Interestingly, Berry points out that the caustics form the only visible elements of geometric optics, since light rays as such are not visible.

Bibliography

[1] J.M. Aarts, *Meetkunde, facetten van de planimetrie en stereometrie*. Epsilon Uitgaven 2000

[2] J.M. Aarts, *Plane and Solid Geometry*. UTM Springer 2008

[3] J.M. Aarts, *Christiaan Huygens: Het Slingeruurwerk. Een studie*. Epsilon Uitgaven **80** 2015

[4] M. Abramowitz and I.A. Stegun, *Handbook of Mathematical Functions with Formulas, Graphs and Mathematical Tables*. National Bureau of Standards, Applied Mathematics Series **55** 1964

[5] K. Andersen, The mathematical technique in Fermat's deduction of the law of refraction. *Historia Mathematica* **10** (1983) 48–62

[6] C. Andriesse, *Titan kan niet slapen, biografie van Christiaan Huygens*. Olympus Pockets 2007

[7] V.I. Arnold, *Mathematical Methods of Classical Mechanics*. GTM **60** Springer 1978; second edition, Springer 1989

[8] V.I. Arnold, *Catastrophe Theory*. Springer 1984

[9] V.I. Arnold, *Singularities of Caustics and Wavefronts*. Mathematics and its Applications (*Soviet Series*) Vol. **62**, Kluwer Academic Publishers 1990

[10] V.I. Arnold, *Huygens & Barrow, Newton & Hooke*. Birkhäuser 1990

[11] V.I. Arnold, On the topological properties of Legendrian projections in the contact geometry of wave fronts. *Algebra and Analysis* **6**(3) (1994) 1–16

[12] V.I. Arnold and A. Avez, *Problèmes Ergodiques de la Mécanique Classique*. Gauthier-Villars 1967

[13] F.M. Arouet (Voltaire), *Candide, ou l'optimisme*, Turtleback 1993 (first edition 1759)

[14] J. Bernoulli, *Opera Johannis Bernoullii* (G. Cramer ed.; 4 vols), Genéve 1742

[15] M.V. Berry, *Nature's Optics and our Understanding of Light*, Contemporary Physics 2015 (to appear)

[16] M.V. Berry and C. Upstill, Catastrophe optics: morphologies of caustics and their diffraction patterns. *Progress in Optics* **18** (1980) 257–346

[17] *BINAS.* Information book for HAVO-VWO in the Natural Sciences. Wolters-Noordhoff 1975

[18] H.J.M. Bos, Lectures in the History of Mathematics. *History of Mathematics* **7** American Mathematical Society – London Mathematical Society 1993

[19] M.G.W. Bos, *De blinde strook.* Master thesis, Rijksuniversiteit Groningen 1982

[20] O. Bottema, Variatierekening. In: L. Kuipers and R. Timman (eds.), *Handboek der Wiskunde 2*, Klassieke wiskunde – Onderwerpen uit de analyse. Scheltema & Holkema 1970

[21] O. Bottema, *Theoretische Mechanica.* Epsilon Uitgaven **3** 1985

[22] H.W. Broer, Huygens' isochrone slinger. *Euclides* **70**(4) (1995) 110–117

[23] H.W. Broer, *Meetkunde en Fysica.* Epsilon Uitgaven **44** 1999

[24] H.W. Broer, *Hemelverschijnselen nabij de Horizon.* Epsilon Uitgaven **77** 2013

[25] H.W. Broer, J. Epema and M. Kuipers, *Oscillaties: trillingen en slingerbewegingen vanuit wiskundig oogpunt.* Internal Publication Rijksuniversiteit Groningen 1987

[26] H.W. Broer and F. Takens, *Dynamical Systems and Chaos.* Epsilon Uitgaven **64** 2009; Appl. Math Sc. **172** Springer 2011

[27] H.W. Broer, Bernoulli's light ray solution of the brachistochrone problem through Hamilton's eyes. *Int. J. Bifurcations Chaos* 24:1440009, 2014

[28] H.W. Broer, Near-horizon celestial phenomena, a study in geometric optics. *Acta Applicandae Mathematicæ* DOI 10.1007/s104400149989-6, 2015

[29] C. Carathéodory, *Gesammelte Mathematische Schriften.* Vols. I–IV, München 1954–7 (Beck) I–V.

[30] M.P. do Carmo, *Differential Geometry of Curves and Surfaces.* Prentice Hall 1976

[31] H.S.M. Coxeter, *Introduction to Geometry.* Wiley 1969

[32] E.D. Dekker (Multatuli), *Max Havelaar, of de koffij-veilingen der Nederlandsche Handel-Maatschappij.* Elsevier 1860; many reprints

[33] E.J. Dijksterhuis, *De Mechanisering van het Wereldbeeld.* Meulenhoff 1950 (third edition 1977); *The Mechanizaton of the World Picture: Pythagoras to Newton.* Princeton University Press 1986

[34] J.J. Duistermaat and W. Eckhaus, *Analyse van Gewone Differentiaalvergelijkingen.* Epsilon Uitgaven **33** 1995

[35] Euclid, *The thirteen books of the Elements.* Three volumes translations and commentary by T.L. Heath, Dover 1956

[36] L. Euler, *Leonhardi Euleri Opera Omnia.* 72 vols., Bern 1911–1975

[37] R.P. Feynman, *QED: The Strange Theory of Light and Matter.* Princeton University Press 1985.

[38] R.P. Feynman, R.B. Leighton and M. Sands, *The Feynman Lectures on Physics* Vols. **1**, **2** and **3**. Addison-Wesley 1963, 1964 and 1965

[39] C. Floor, De ondergaande zon. *Zenit* **4** (1978) 126–128

[40] C. Floor, Atmosferische straalkromming. *Natuur en Techniek* juli 1978

[41] C. Floor, De vorm van de zonneschijf bij lage zonnestand, computersimulaties en waarnemingen. *Meteorological* maart en april 1999

[42] C.G. Fraser, Sufficient conditions, fields and the calculus of variations. *Historia Mathematica* (2009), DOI 10.1016/j.hm.2009.02.001

[43] H. Goldstein, *Classical Mechanics.* Addison-Wesley 1950 (second edition 1980)

[44] H.H. Goldstine, *A History of the Calculus of Variations from the 17th through the 19th Century.* Studies in the History of Mathematics and Physical Sciences **5** Springer 1980

[45] M. Golubitsky, An introduction to catastrophe theory and its applications. *SIAM Review* **20**(2) (1978), 352–387

[46] V. Guillemin and S. Sternberg, *Geometric Asymptotics.* Mathematical Surveys **14** Americal Mathematical Society 1977

[47] W.R. Hamilton, Theory of systems of rays. *Trans. Roy. Irish Acad.* **15** (1828), 69–174

[48] W.R. Hamilton, On a general method in dynamics. *Phil. Trans. Roy. Soc.* **124** (1834), 247–308

[49] W.R. Hamilton, Second essay on a general method in dynamics. *Phil. Trans. Roy. Soc.* **125** (1835), 95–144

[50] K.P. Hart, De Cycloïde. *Pythagoras* **39**(4) 2000

[51] A. van Helden, *Measuring the Universe, Cosmic Dimensions from Aristarchus to Halley.* The University of Chicago Press 1985

[52] M.W. Hirsch, *Differential Topology.* Springer 1976

[53] M.W. Hirsch, S. Smale and R.L. Devaney, *Differential Equations, Dynamical Systems, and an Introduction to Chaos.* Academic Press 2004

[54] W.J. Humphries, *Physics of the Air.* Franklin Institute 1920; third edition McGraw-Hill 1940; Dover 1965

[55] Chr. Huygens, *Œvres Complètes de Christiaan Huygens publiées par la Société Hollandaise des Sciences* **16**, Martinus Nijhoff, The Hague 1929 Vol. 5, 241–262; Vol. 17, 156–189

[56] Chr. Huygens, *Verhandeling over het Licht.* Epsilon Uitgaven **18** 1990

[57] J. Kepler, *Ad Vitellium Paralipomena, quibus Astronimiae pars optica traditur.*, Claude Marne and Erven Johannes Auber, (Frankfurt) 1604; In: W. von Dyck and M. Caspar. *Gesammelte Werke band* II, C.H. Becksche Verlagsbuchsabhandlungen (München) 1938

[58] D.E. Kirk, *Optimal Control Theory: an Introduction.* Dover 2004

[59] A. Koestler, *The Sleepwalkers.* Hutchinson 1959; Danube 1979

[60] J.L. Lagrange, *Mécanique Analytique.* 4 ed., 2 vols. Paris: Gauthier-Villars et fils 1888–89

[61] J.L. Lagrange, *Œuvres.* A. Serret and G. Darboux eds., Paris 1867–1892

[62] H. de Lang, Het principe van de minste actie: I. Klassiek. *Nederl. Tijdschr. voor Natuurkunde* **78**(9) (2012) 362–366

[63] H. de Lang, Het principe van de minste actie: II. Quantummechanisch. *Nederl. Tijdschr. voor Natuurkunde* **78**(10) (2012) 390–393

[64] H. de Lang, Het principe van de minste actie: III. Van padintegraal tot Feynmandiagrammen. *Nederl. Tijdschr. voor Natuurkunde* **78**(11) (2012) 420–425

[65] W.H. Lehn, Inversion of superior mirage data to compute temperature profiles. *J. Opt. Soc. Am.* **73**(12) (1983) 1622–1625

[66] G.W. Leibniz, *Nova methodus pro maximis et minimis. Acta Eruditorum* 1684; In: D.J. Struik, *A Source Book in Mathematics* 1200–1800. Harvard University Press 1969 271–281

[67] U. Leonhardt and Th. Philbin, *Geometry and Light: the science of invisibility.* Dover 2010

[68] M. Levi, *The Mathematical Mechanic using physical: reasoning to solve problems.* Princeton University Press 2009

[69] C.M. Linton, *From Eudoxus to Einstein, A History of Mathematical Astronomy.* Cambridge University Press 2004

[70] J.A. van Maanen, *Een Complexe Grootheid, leven en werk van Johann Bernoulli,* 1667–1748. Epsilon Uitgaven **34** 1995

[71] J. Martin, The Helen of Geometry. *The College Mathematical Journal* **41** (2010) 17–28

[72] M.G.J. Minnaert, *De Natuurkunde van 't Vrije Veld* delen **1**, **2** en **3**, Vijfde Editie, ThiemeMeulenhoff 1996; *The Nature of Light and Colour in the Open Air*, Dover 1954 (first edition 1937–40)

[73] L. Molenaar, *De Rok van het Universum – Marcel Minnaert astrofysicus* 1893–1970. Uitgeverij Balans 2003

[74] I. Newton, *Philosophiæ Naturalis Principia Mathematica.* Translated by I. Bernard Cohen and Anne Whitman as *Isaac Newton, The Principia – Mathematical Principles of Natural Philosophy.* University of California Press 1999 (first edition 1687)

[75] I. Newton, *Opticks.* Dover 1952 (first edition 1704)

[76] E. Noether, Invariante Variationsprobleme. *Nachrichten der Königlichen Gesellschaft der Wissenschaften zu Göttingen, Math-phys. Klasse* (1918), 235–257

[77] D.J.K. O'Connell, *The Green Flash and Other Low Sun Phenomena.* North-Holland 1958

[78] F.W.J. Olver, D.W. Lozier, R.F. Boisvert, C.W. Clark (eds.), *NIST Handbook of Mathematical Functions.* Cambridge University Press 2010

[79] A.D. Pierce, *Acoustics: an Introduction to its Physical Principles and Applications.* McGraw-Hill Book Company, Inc., New York, 1981. Also available from the Acoustical Society of America.

[80] J.P. Philips, Brachistochrone, Tautochrone, Cycloid – Apple of Discord. *The Mathematics Teacher* **60**(5) (1967) 506–508

[81] T. Poston and I. Stewart, *Catastrophe Theory and its Applications.* Pitman 1978

[82] G.F.B. Riemann, Über die Hypothesen, welche der Geometrie zu Grunde liegen (Habilitationsschrift, 1854). *Abhandlungen der Königlichen Gesellschaft der Wissenschaften zu Göttingen* **13** (1868)

[83] A.I. Sabra, *Theories of Light from Descartes to Newton.* Osbourne 1967

[84] H. Seifert and W. Threlfal, *Variationsrechnung im Grossen.* Chelsea Publishing Company 1938

[85] D.G. Shafer, The brachistochrone: historical gateway to the calculus of variations. *MATerials MATemàtics* **5**(14) 2007

[86] D.E. Smith, *A Source Book in Mathematics.* McGraw-Hill Book Company 1929; Dover 1959

[87] D. Sobel, *Longitude.* Walker 1995; 2005

[88] M. Spivak, *Differential Geometry.* Vol. **I** Publish or Perish 1970

[89] A.N. Strahler, *Physical Geography.* Wiley 1951; 1960; 1969; 1975

[90] D.J. Struik, *A Source Book in Mathematics* 1200–1800. Harvard University Press 1969

[91] H.J. Sussmann and J.C. Willems, 300 years of optimal control from the brachystochrone to the maximum principle. *IEEE Control Systems* (1997) 32–44

[92] J. Verne, *Le Rayon Vert.* Hetzel (Parijs) 1881

[93] A.L. Wegener, Über die Ursache der Zerrbilder bei Sonnenuntergängen. *Beiträge zur Physik der freien Atmosphäre* Strasbourg, **4** (1912) 26–34

[94] A.L. Wegener, Optik der Atmosphäre. In: Müller–Pouillet, *Lehrbuch der Physik*, 2nd edn., Viewig & Son, Braunschweig (1928) 266–289

[95] S.Y. van der Werf, *Het Nova Zembla Verschijnsel; geschiedenis van een luchtspiegeling.* Historische Uitgeverij 2011

[96] S.Y. van der Werf, G.P. Können, W.H. Lehn and F. Steenhuizen, Waerachtige beschrijvingen van het Nova Zembla effect. *Nederlands Tijdschrift voor Natuurkunde* **66** (2000) 120–126

[97] E.A. Whitman, Some historical notes on the cycloid. *The American Mathematical Monthly* **50**(5) (1943) 309–315

[98] J.G. Yoder, *Unrolling Time.* Cambridge University Press 1988

Index